Constant Mean Curvature Immersions of Enneper Type

Recent Titles in This Series

(Continued in the back of this publication)

MEMOIRS
of the
American Mathematical Society

Number 478

Constant Mean Curvature Immersions of Enneper Type

Henry C. Wente

November 1992 • Volume 100 • Number 478 (first of 4 numbers) • ISSN 0065-9266

American Mathematical Society
Providence, Rhode Island

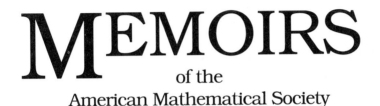

1991 *Mathematics Subject Classification.*
Primary 53A10.

Library of Congress Cataloging-in-Publication Data

Wente, Henry C., 1936–
Constant mean curvature immersions of Enneper type/Henry C. Wente.
p. cm. – (Memoirs of the American Mathematical Society, ISSN 0065-9266; no. 478)
Includes bibliographical references.
ISBN 0-8218-2536-4
1. Minimal surfaces. I. Title. II. Series.
QA3.A57 no. 478
[QA644]
510 s–dc20 92-28574
[516.3′62] CIP

Memoirs of the American Mathematical Society

This journal is devoted entirely to research in pure and applied mathematics.

Subscription information. The 1992 subscription begins with Number 459 and consists of six mailings, each containing one or more numbers. Subscription prices for 1992 are $292 list, $234 institutional member. A late charge of 10% of the subscription price will be imposed on orders received from nonmembers after January 1 of the subscription year. Subscribers outside the United States and India must pay a postage surcharge of $25; subscribers in India must pay a postage surcharge of $43. Expedited delivery to destinations in North America $30; elsewhere $82. Each number may be ordered separately; *please specify number* when ordering an individual number. For prices and titles of recently released numbers, see the New Publications sections of the *Notices of the American Mathematical Society.*

Back number information. For back issues see the *AMS Catalogue of Publications.*

Subscriptions and orders should be addressed to the American Mathematical Society, P. O. Box 1571, Annex Station, Providence, RI 02901-1571. *All orders must be accompanied by payment.* Other correspondence should be addressed to Box 6248, Providence, RI 02940-6248.

Memoirs of the American Mathematical Society is published bimonthly (each volume consisting usually of more than one number) by the American Mathematical Society at 201 Charles Street, Providence, RI 02904-2213. Second-class postage paid at Providence, Rhode Island. Postmaster: Send address changes to Memoirs, American Mathematical Society, P. O. Box 6248, Providence, RI 02940-6248.

Printed in the United States of America.
This volume was printed directly from author-prepared copy.
The paper used in this book is acid-free and falls within the guidelines established to ensure permanence and durability. ♾
10 9 8 7 6 5 4 3 2 1 97 96 95 94 93 92

TABLE OF CONTENTS

ABSTRACT

This paper considers surfaces of constant mean curvature H immersed in Euclidean space R^3 (or more generally in $M^3(c)$ the simply connected spaces of constant curvature c) with the property that one family of curvature lines are spherical. Such surfaces are said to be of Enneper type. This geometrical problem is reduced to finding solutions to the Gauss equation $\Delta\omega + (H^2 + c)e^{2\omega} - Be^{-2\omega} = 0$ where B is a positive constant and such that $\omega(u,v)$ satisfies the auxillary condition $2\omega_u = \alpha(u)e^{\omega} + \beta(u)e^{-\omega}$ for suitable functions $\alpha(u),\beta(u)$. The nature of the solutions depend strongly on the sign of $H^2 + c$.

Key words: constant mean curvature immersions, spherical lines of curvature, elliptic Sinh-Gordon equation.

Note: Work on this paper was done in part while the author was a guest of SFB 256 at University of Bonn, Germany and also with the support of the National Science Foundation.

The author would like to express his appreciation to Ivan Sterling for providing the computer-generated illustrations appearing at the end of this paper and to Pamela Zawierucha for typing and final preparation of this work.

Received by Editors December 23, 1990.

I. INTRODUCTION

In this paper we carry out the construction of constant mean curvature (cmc) immersed surfaces in R^3 (or more generally the spaces $M^3(c)$ of constant curvature c) which satisfy the following geometric condition. One family of curvature lines of the immersed surface is spherical; that is, each curvature line of the family is to lie on some sphere. Immersed surfaces satisfying this condition have been the subject of considerable study by classical differential geometers. An extensive discussion may be found in the treatise of G. Darboux [4], see also L.P.Eisenhart [7], and there was a book on the subject by A. Enneper [8] in 1880. For this reason we shall call any surface satisfying this geometric condition a surface of Enneper type.

Let x(u,v) be a cmc immersion from an open set in R^2 into $M^3(c)$. Suppose that the immersion is conformal and that the preimage of the lines of curvature are straight lines parallel to the coordinate axes. If we write the first fundamental form

$$(1.1) \qquad ds^2 = e^{2\omega}(du^2 + dv^2)$$

then, as will be developed in Section II, the Gauss equation will be

$$(1.2) \qquad \Delta\omega + Ae^{2\omega} - Be^{-2\omega} = 0.$$

Here B is a positive constant while $A = H^2 + c$ where H is the mean curvature of the immersion. The character of the solution is strongly influenced by the sign of A so we shall consider each of these possibilities: $H = 1/2$ in R^3, minimal surfaces $H = 0$ in R^3, and minimal surfaces $H = 0$ in hyperbolic 3-space $M^3(-1) \equiv H^3$. If the immersion is of Enneper type then the following auxilliary condition must be satisfied

$$(1.3) \qquad 2\omega_u = \alpha(u)e^{\omega} + \beta(u)e^{-\omega}$$

where (assuming that $\omega_v \not\equiv 0$) $\alpha(u)$, $\beta(u)$ will be solutions to a second order system of differential equations from which the solution $\omega(u,v)$ is to be recovered. The system is

$$\alpha'' = a\alpha - 2\alpha^2\beta - 2A\beta$$
(1.4)
$$\beta'' = a\beta - 2\alpha\beta^2 - 2B\alpha \qquad a = \text{constant}.$$

Furthermore one finds

$$(1.5)\quad 4\omega_v^{\,2} = -(4A + \alpha^2)e^{2\omega} - (4B + \beta^2)e^{-2\omega} - 4\alpha' e^{\omega} + 4\beta' e^{-\omega} + 6\gamma$$

where $6\gamma = 6\alpha\beta - 4a$.

If one solves the system (1.4) then (1.3) and (1.5) may be used to recover $\omega(u,v)$. This development is carried out in Section II. The system (1.4) is an algebraic completely integrable Hamiltonian system. To solve it we follow the method described in Darboux [4] which is to solve the relevant Hamilton-Jacobi equation by the method of separation of variables. A general feature of any solution $\omega(u,v)$ obtained in this manner is that it will be periodic in the v-direction (with infinite period allowed) and quasi-periodic in the u-direction.

In section III we study the case with mean curvature $H = 1/2$ in R^3. Here the Gauss Equation (1.2) may be written in the form

$$(1.6)\qquad \Delta\omega + \sinh\omega\cosh\omega = 0, \qquad A = B = 1/4.$$

This is the P.D.E. which initially attracted my attention when constructing immersed cmc tori in R^3 [13]. One finds here that if an immersion of Enneper type is defined locally then it extends to a global mapping of the plane into R^3. In particular the immersion will not develop any umbilic points. One can explicitly compute the centers and radii of the spheres determined by the lines of curvature from the solutions to the system (1.4). It turns out that any immersion has an axis ℓ on which the centers of all spheres lie. This fact enables one to give fairly explicit formulae for the immersion itself.

All solutions to the elliptic sinh-Gordon equation (1.6) which are doubly periodic have been classified recently in a remarkable paper by U. Pinkall and I. Sterling [11]. These solutions fall into classes indexed by the positive integers. The solutions constructed here lie in the class N = 2. Solutions of finite type are also amenable to the techniques of soliton theory. This aspect has recently been carried out by A.I. Bobenko [2].

In Section III we also look at some specific interesting examples. For any integer $n \geq 2$ one can construct an immersed cylinder with two embedded ends which are asymptotically round cylinders and which has n sphere-like lobes attached in the middle. It is symmetric about a plane perpendicular to the axis passing through these lobes. This striking surface seems to have been first observed by Pinkall and Sterling and a computer picture of it appears in their paper [11]. Instead of embedded ends which are round cylinders one can also construct surfaces whose embedded ends are asymptotic to Delaunay unduloids of any type. Again these cmc immersions will have n-lobes in the middle. These immersions now come in one-parameter families. Also some of the complicated quasi-periodic examples illustrated in [11] seem to be of Enneper type.

One finds many possibilities of immersed tori of Enneper type. In fact, one expects that they should fall into one-parameter families, and one verifies that this is true at least in certain cases.

In Section IV we look at the case of minimal surfaces in R^3. This problem was solved by H. Dobriner [6] in 1887 exhibiting the possible minimal surfaces using theta functions. For the sake of completeness I rederive the results here. The Gauss equation is now the classical Liouville equation. The solution $\omega(u,v)$ constructed here will have the same general features as for the case $H = 1/2$ except that $\omega(u,v)$ will have scattered in the plane points where the solution becomes positively infinite. These correspond to flat ends for the immersed minimal surface. In particular, we look for the Weierstrass representation of these surfaces which is expressed explicitly in terms of the Weierstrass $\wp$-function. The surfaces depicted here resemble a catenoid, perhaps covered infinitely often, from which a number of flat ends have been extruded.

In Section V we consider the case of minimal immersions into hyperbolic 3-space $M^3(-1) \equiv H^3$. This leads to the Gauss equation

(1.7) $$\Delta\omega = e^{2\omega} + e^{-2\omega}$$

The techniques developed in Section II still apply although now the separation of variables method becomes more complicated. Again the solutions will be periodic in the v-direction. Now the points of singularity which developed for minimal surfaces in R^3 have expanded into holes. The maximal domain for the solutions to (1.7) is no longer simply connected. It will be a planar domain with periodic holes. Also in general the domain will be contained inside a vertical strip. The lines of curvature in the v-direction must now lie on spheres, horospheres, or "pseudo-" spheres. Once again the centers of the spheres all lie on a geodesic line in H^3. We also look at some examples. In particular, one finds examples where the immersion has ends which are the H^3 counterpart of Delaunay surfaces.

We conclude the introduction with the following historical remarks. The simplest geometrical condition one can impose on a cmc surface in R^3 is that it be a surface of revolution. In this situation one obtains the classical Delaunay surfaces.

A Joachimsthal surface is one for which the lines of curvature of one family are all planar and such that these planes all contain a common line, the axis. It follows that the lines of curvature of the other family are all spherical with the centers of the spheres lying on the axis. Furthermore, the immersed surface will intersect these spheres at right angles along these curvature lines (see Eisenhart [7]).

If x(u,v) is an H = 1/2 immersion in R^3 with planar lines of curvature in the u-direction then the parallel surface $\mathbf{y} = \mathbf{x} + \xi$ where ξ is the oriented unit normal vector, is an immersed surface of constant Gauss curvature K = +1. It is a Joachimsthal surface. For this reason we call the corresponding cmc immersion a surface of Joachimsthal type. It is a special case of an Enneper type surface. These were the surfaces initially studied by the author in the construction of closed cmc tori in R^3, [14]. The explicit representation of these surfaces in terms of elliptic integrals has been carried out by U. Abresch [1] and also by R. Walter [13]. We should also note that the Joachimsthal surfaces with K = -1 are treated in Darboux [4] (see also Eisenhart [7]). In this case one wants to solve the hyperbolic sine-Gordon equation. The treatment in both cases is similar.

Finally, H. Dobriner [5] studied the case of Enneper type surfaces of constant negative Gauss curvature $K = -1$. His treatment is based on the book of Enneper [8] and expresses the immersions using theta functions. His work is discussed in Darboux [4]. The analysis in the present work follows the approach in Darboux's treatise.

If a minimal surface in R^3 has one family of planar curvature lines then the same must be true of the other family. Besides the catenoid and Enneper's minimal surface there is a one-parameter family of such surfaces called Bonnet's surfaces. These are described in Eisenhart [7]. In some sense these surfaces serve to connect the catenoid to Enneper's minimal surface. Finally, in Section V we shall find that surfaces of revolution are the only Joachimsthal type minimal surface in hyperbolic space, a somewhat surprising result.

II. THE DIFFERENTIAL GEOMETRY

The construction of cmc immersions is based on the following well-known result due to O. Bonnet in the Euclidean case [12], which we state as a theorem.

<u>Theorem 2.1</u>: Let $x = F(w)$, $w = u + iv$ be a conformal cmc immersion from an open set in R^2 into $M^3(c)$, the complete simply-connected Riemannian 3-manifold of constant curvature c. Suppose the first and second fundamental forms are given by

$$\text{(2.1)} \quad \begin{aligned} &\text{a)} \quad ds^2 = e^{2\omega}(du^2 + dv^2) \\ &\text{b)} \quad -(dx \cdot d\xi) = Ldu^2 + 2Mdudv + Ndv^2. \end{aligned}$$

The Codazzi-Mainardi equations are equivalent to

$$\text{(2.2)} \qquad \phi(w) = \left(\frac{L - N}{2}\right) - i\, M$$

is holomorphic, while the Gauss equation becomes

$$\text{(2.3)} \qquad \Delta\omega + (H^2 + c)e^{2\omega} - |\phi|^2 e^{-2\omega} = 0$$

where H is the mean curvature of the immersion. Conversely, given the pair $\{\phi(w), \omega(u,v)\}$ satisfying (2.2) (2.3) on a simply connected region Ω there is determined a cmc immersed surface $x(u,v)$ from Ω into $M^3(c)$ with mean curvature H whose fundamental forms are as above. This immersion is unique to within an isometry of $M^3(c)$.

A straightforward calculation establishes that

$$\text{(2.4)} \qquad |\phi(w)| = |k_1 - k_2| e^{2\omega}/2$$

where k_1, k_2 are the principal curvatures. Given that $k_1 + k_2 = 2H$ we find

$$(2.5)\qquad k_1 = H - |\phi|e^{-2\omega}, \quad k_2 = H + |\phi|e^{-2\omega}$$

assuming that $k_1 \le k_2$. the extrinsic Gauss curvature is defined to be $K_e = k_1 k_2$ and the intrinsic Gauss curvature

$$(2.6)\qquad K = K_e + c = (H^2 + c) - |\phi|^2 e^{-4\omega}$$

Note: If a new set of coordinates is introduced by an analytic mapping $w = f(w')$ then one obtains a new pair $\{\hat{\phi},\hat{\omega}\}$. They are related to the original pair by

$$(2.7)\qquad \begin{aligned} &a)\quad e^{2\hat{\omega}} = e^{2\omega}|f'(w')|^2 \\ &b)\quad \hat{\phi}(w') = \phi(w)\cdot f'(w')^2 \end{aligned}$$

From the latter equation we see that

$$(2.8)\qquad \hat{\phi}(w')dw'^2 = \phi(w)dw^2$$

This is the Hopf quadradric diferential [9].

From (2.4) we see that the zeros of $\phi(w)$ correspond to the umbilic points of the immersion. If $\phi(w) \equiv 0$ then all points are umbilic and the immersed surface is a sphere. Otherwise $\phi(w) \not\equiv 0$ and the umbilic points are isolated. For a simply connected region where the Hopf differential does not vanish one can make a conformal change of coordinates so that $\phi(w) = \text{constant} \not\equiv 0$. The lines of curvature are given by

$$(2.9)\qquad \mathrm{Im}[\phi(w)dw^2] = -Mdu^2 + (L - N)dudv + Mdv^2 = 0$$

from which it follows that the preimage of the lines of curvature are straight lines in the parameter domain. By a rotation one can make $M = 0$, $\phi(w)$ a real constant, and the lines of curvature parallel to the coordinate axes.

For a cmc immersion into R^3 with H positive we may set $\phi(w) = -H$. This implies

$$k_1 = 2He^{-\omega}\sinh\omega, \qquad k_2 = 2He^{-\omega}\cosh\omega$$

(2.10)

$$-dx\cdot d\xi = 2He^{\omega}\,(\sinh\omega\, du^2 + \cosh\omega\, dv^2).$$

The k_1–curvature lines are parallel to the u-axis. Setting $\phi(w) = H$ would put the k_2-curvature lines parallel to the u-axis. The Gauss equation (2.3) becomes

$$\Delta\omega + H^2(e^{2\omega} - e^{-2\omega}) = 0$$

(2.11)

$$\Delta\omega + 4H^2\sinh\omega\cosh\omega = 0$$

From the form of (2.11) it is convenient to choose $H = 1/2$.

For the case of minimal immersions into R^3 we select $\phi(w) = -1$. This makes

(2.12)
$$k_1 = -e^{-2\omega}, \qquad k_2 = e^{-2\omega}$$

$$-\,(dx\cdot d\xi) = -\,du^2 + dv^2$$

and the Gauss equation is now

(2.13)
$$\Delta\omega - e^{-2\omega} = 0$$

which is the classical Liouville equation.

Finally, for cmc immersions with $0 \le H < 1$ into hyperbolic space $M^3(-1)$ we select $\phi(w) = -\sqrt{1 - H^2}$. This makes

$$k_1 = H - \sqrt{1 - H^2}\, e^{-2\omega}, \quad k_2 = H + \sqrt{1 - H^2}\, e^{-2\omega}$$

(2.14)

$$-dx\cdot d\xi = (He^{2\omega} - \sqrt{1 - H^2})du^2 + (He^{2\omega} + \sqrt{1 - H^2})dv^2$$

and the Gauss equation becomes

(2.15)
$$\Delta\omega = (1 - H^2)(e^{2\omega} + e^{-2\omega}).$$

For convenience we shall set $H = 0$.

We now impose the condition that the immersion be of Enneper type. Specifically we demand that the lines of curvature corresponding to the parameter lines u = constant be spherical.

<u>Theorem 2.2</u>: Suppose x(u,v) is a conformal cmc immersion into $M^3(c)$ of Enneper type parameterized by its lines of curvature curvature such that the lines of curvature in the v-direction are spherical and with fundamental forms (2.1). Besides satisfying the Gauss equation

$$\Delta\omega + Ae^{2\omega} - Be^{-2\omega} = 0 \tag{2.16}$$

a) For H = 1/2 in R^3, A = B = 1/4

b) For H = 0 in R^3, A = 0, B = 1

c) For H = 0 in H^3, A = -1, B = 1

there are functions $\alpha(u)$, $\beta(u)$ such that

$$2\omega_u = \alpha(u)e^{\omega} + \beta(u)e^{-\omega} . \tag{2.17}$$

<u>Proof</u>: To be discussed in Sections III and IV.

We now want to solve the system (2.16)(2.17). A key step is the following result.

<u>Theorem 2.3</u>: Let $\omega(u,v)$ be a solution to the system (2.16 - 17)

$$\Delta\omega + Ae^{2\omega} - Be^{-2\omega} = 0$$

$$2\omega_u = \alpha(u)e^{\omega} + \beta(u)e^{-\omega}.$$

If $\omega_v \not\equiv 0$ then the functions $\{\alpha(u), \beta(u)\}$ are solutions to the system

$$\begin{aligned} \alpha'' &= a\alpha - 2\alpha^2\beta - 2A\beta \\ \beta'' &= a\beta - 2\alpha\beta^2 - 2B\alpha \end{aligned} \tag{2.18}$$

where a is a constant. Furthermore

$$(2.19)\quad 4\omega_v^{\,2} = -(4A + \alpha^2)e^{2\omega} - (4B + \beta^2)e^{-2\omega} - 4\alpha' e^{\omega} + 4\beta' e^{-\omega} + 6\gamma$$

where $6\gamma = 6\alpha\beta - 4a$. The system (2.18) is an algebraic completely integrable Hamiltonian system with

$$\mathcal{H}(\alpha, \beta, p_\alpha, p_\beta) = p_\alpha p_\beta - \psi(\alpha,\beta) = h$$

$$\psi(\alpha,\beta) = a\alpha\beta - \alpha^2\beta^2 - A\beta^2 - B\alpha^2.$$

Conversely, let $\{\alpha(u),\beta(u)\}$ be a solution to(2.18) with the property that the right side of (2.19) is positive for $u = u_o$ and some $\omega = \omega_o$. One may use (2.19) to solve for $\omega(u_o,v)$ and then (2.17) to find $\omega(u,v)$.

Proof: Let $\omega(u,v)$ be a solution to (2.16) with $\omega_v \not\equiv 0$ and such that there are functions $\{\alpha(u), \beta(u)\}$ satisfying (2.17). We first derive (2.19). Compute ω_{uu} by differentiating (2.17) and substitute this into (2.16) to obtain an expression for $4\omega_{vv}$. Multiply this by $2\omega_v$ and integrate in v to obtain (2.19) where $\gamma(u)$ is some function of u. Now compute $8\omega_v\omega_{uv}$ in two ways, once by differentiating (2.19) with respect to u and also by differentiating (2.17) with respect to v and multiplying the result by $4\omega_v$. We obtain a polynomial expression in $X = e^{\omega}$ with coefficients being functions of u which is equal to zero. This gives us the system (2.18) and the condition that $6\gamma(u) = 6\alpha\beta - 4a$. The converse is just a reversal of the procedure. Q.E.D.

The solutions $\omega(u,v)$ obtained in this manner have some specific properties which we now list.

1) By setting $X = e^{\omega}$ we may rewrite (2.19) in the form

$$(2.20)\qquad 4X_v^{\,2} = p(u,X)$$

where $p(u,X) = -(4A + \alpha^2)X^4 - 4\alpha' X^3 + 6\gamma X^2 + 4\beta' X - (4B + \beta^2)$. Here $p(u,X)$ is a fourth degree polynomial in X with the coefficients functions of u. It is necessary that $p(u,X)$ be positive for some positive values of $X = e^{\omega}$ in order for a solution $\omega(u,v)$ to exist.

A fourth degree polynomial has two invariants I and J defined as follows. If

$$p(x) = a_0 x^4 + 4a_1 x^3 + 6a_2 x^2 + 4a_3 x + a_4$$

$$I = a_0 a_4 - 4a_1 a_3 + 3a_2^2 \tag{2.21}$$

$$J = \det \begin{bmatrix} a_0 & a_1 & a_2 \\ a_1 & a_2 & a_3 \\ a_2 & a_3 & a_4 \end{bmatrix}$$

while the discriminant $\Delta = I^3 - 27J^2$. A direct calculation shows that the invariants associated with the polynomials $p(u,X)$ in (2.20) are constant and independent of u (i.e. they are integrals of the system (2.18)).

2) Suppose that for $u = u_0$ we have $p(u_0,X)$ in (2.20) positive for some $X_0 > 0$. We may use (2.20) to find $\omega(u_0,v)$ and then (2.17) to find $\omega(u,v)$. We observe that (2.17) is a Ricatti equation in $X = e^{\omega}$.

$$2X_u = \alpha(u)X^2 + \beta(u). \tag{2.22}$$

3) If $\omega_v(u_0,B) = 0$ then $\omega(u,v)$ is symmetric about the line $v = B$ satisfying

$$\omega_v(u,B) = 0$$

$$\omega(u,B + v) = \omega(u,B - v).$$

This follows from the construction procedure indicated in (2).
4) There are in general six parameters, the constant a in (2.18), the initial conditions $\alpha(u_o)$, $\beta(u_o)$, $\alpha'(u_o)$, $\beta'(u_o)$ and $\omega(u_o,v_o)$. Taking into account the translation of coordinates, we obtain a four-parameter family of distinct solutions to (2.16).

We shall solve the system (2.18) by the method of separation of variables applied to the Hamiltonian-Jacobi equation. We look for a solution $\theta(\alpha,\beta,h,k)$ to the P.D.E.

(2.23) $$\theta_\alpha \theta_\beta = a\alpha\beta - \alpha^2\beta^2 - A\beta^2 - B\alpha^2 + h$$

Make the change of variables

(2.24) $$\begin{aligned} \alpha\beta &= s + t - 4\sqrt{AB} \\ (\sqrt{B}\,\alpha - \sqrt{A}\,\beta)^2 &= -\,st. \end{aligned}$$

With respect to these new variables the Hamilton-Jacobi equation becomes

(2.25) $$\begin{aligned} s(s - 4\sqrt{AB})\theta_s^{\,2} - t(t - 4\sqrt{AB})\theta_t^{\,2} &= g(s) - g(t) \\ g(s) = -\,s^3 + (a + 6\sqrt{AB})s^2 &+ (h - 4a\sqrt{AB} - 8AB)s + k. \end{aligned}$$

Writing $\theta(s,t) = P(s) + Q(t)$ separates the variables allowing us to find θ by quadratures. We differentiate with respect to the parameters h, k setting $\theta_h = u - u_o$ and $\theta_k = k'$ where k' is a new parameter. This determines s and t. In fact, one obtains the following recipe. Let $s(\lambda)$, $t(\lambda)$ be distinct solutions to

(2.26) $$\begin{aligned} s'(\lambda)^2 &= s(s - 4\sqrt{AB})g(s) \\ t'(\lambda)^2 &= t(t - 4\sqrt{AB})\,g(t) \end{aligned}$$

where $\lambda = \lambda(u)$ is determined by the condition

(2.27) $$\lambda'(u) = \frac{2}{s - t}.$$

In theory we have found s and t as functions of u giving us the solutions $\{\alpha(u),\beta(u)\}$. This is the classical procedure of Jacobi.

Suppose that $A \geq 0$ and B is positive. In this case the transformation (2.24) is real and we find

$$\begin{aligned}(\sqrt{B}\,\alpha - \sqrt{A}\,\beta)^2 &= -\,st\\ (\sqrt{B}\,\alpha + \sqrt{A}\,\beta)^2 &= -(s - 4\sqrt{AB})(t - 4\sqrt{AB}).\end{aligned} \tag{2.28}$$

In the transformation either $s > 4\sqrt{AB}$ and $t < 0$ or conversely. Therefore in solving (2.26) we look for solutions $\{s(\lambda),\ t(\lambda)\}$ satisfying $s > 4\sqrt{AB}$ and $t < 0$.

There is an ambiguity in recovering $\langle\alpha,\beta\rangle$ from $\langle s,t\rangle$. If $\langle\alpha(u),\ \beta(u)\rangle$ satisfies (2.24) then so does $\langle-\alpha(u),\ -\beta(u)\rangle$. The proper choice becomes important in (2.20). It is necessary to choose $\langle\alpha,\ \beta\rangle$ so that the polynomial $p(u,\ X)$is positive for some positive X. If $p(u,X)$ is the polynomial corresponding to $\langle\alpha,\beta\rangle$ then $p(u,-X)$ is the polynomial corresponding to $\langle-\alpha,\ -\beta\rangle$.

One may start with the polynomial $g(s)$ given by (2.25) where it is necessary that $g(s)$ be positive for some $s > 4\sqrt{AB}$. Write $g(s)$ in the form

$$\begin{aligned}4\,g(s) &= 4(-s + p)^3 - I(-s + p) + J\\ p &= \frac{a + 6\sqrt{AB}}{3}\end{aligned} \tag{2.29}$$

Then

$$\begin{aligned}I &= 16AB + 4h + \frac{4}{3}a^2\\ J &= 4k + \frac{4h}{3}(a + 6\sqrt{AB}) + \frac{8}{27}a^3 - \frac{32}{3}aAB\end{aligned} \tag{2.30}$$

where these are the invariants of the polynomial $p(u,\ X)$ appearing in (2.20). In particular one finds that

$$4k = (\alpha\beta' - \alpha'\beta)^2 + 4(\sqrt{A}\beta' - \sqrt{B}\alpha')^2 + 4(\alpha\beta - a - 2\sqrt{AB})(\sqrt{B}\alpha - \sqrt{A}\beta)^2 \tag{2.31}$$

where k is the constant term in the polynomial $g(s)$ in (2.25).

III. H = 1/2 IMMERSIONS IN R^3

Let x(u,v) be a conformal immersion parameterized by the lines of curvature with fundamental forms and Gauss equation

$$ds^2 = e^{2\omega}(du^2 + dv^2)$$

$$\text{(3.1)} \qquad II = e^{\omega}(\sinh\omega\, du^2 + \cosh\omega\, dv^2)$$

$$III = \sinh^2\omega\, du^2 + \cosh^2\omega\, dv^2$$

with principal curvatures $k_1 = e^{-\omega}\sinh\omega$ and $k_2 = e^{-\omega}\cosh\omega$

$$\text{(3.2)} \qquad \Delta\omega + \sinh\omega \cosh\omega = 0.$$

The Enneper property (2.17) is a consequence of the following result (see Eisenhart [7]).

<u>Lemma 3.1</u>: Let x(u,v) be an immersed surface in R^3 parameterized by the lines of curvature with fundamental forms

$$ds^2 = Edu^2 + Gdv^2$$

$$II = Ldu^2 + Ndv^2$$

$$III = \mathfrak{E}du^2 + \mathfrak{G}dv^2.$$

Suppose that the immersion is of Enneper type such that for each $\hat{u}$ the image $x(\hat{u},v)$ lies on the sphere $S[p(\hat{u}),R(\hat{u})]$ with center $p(\hat{u})$ and radius $R(\hat{u})$. Let $e_1 = x_u/|x_u|$, $e_2 = x_v/|x_v|$, $e_3 = \xi$ be the orthonormal Darboux frame and set

$$\text{(3.3)} \qquad p(\hat{u}) - x(\hat{u},v) = (R\sin\theta)e_1 + (R\cos\theta)e_3 \equiv x_o e_1 + z_o e_3$$

where x_o and z_o are functions of u alone. Here $\Theta(\hat{u})$ is the angle of intersection of the surface with the sphere along the line of curvature $x(\hat{u},v)$. We have

$$\rho_2 = \varepsilon x_o[(\sqrt{\mathcal{G}})_u/\sqrt{\mathcal{E}\mathcal{G}}] + z_o, \qquad \varepsilon = -\text{sign}(k_1) \tag{3.4}$$

where ρ_2 is the radius of curvature in the v-direction.

If we apply Lemma 3.1 to our cmc immersions we find

$$\omega_u = y(u)\cosh\omega + z(u)\sinh\omega$$
$$\tan\Theta = \frac{1}{y - z}, \qquad R^2 = \frac{1 + (y - z)^2}{z^2}. \tag{3.5}$$

Condition (3.5) is identical to (2.17) where in this section it is more convenient to use the functions y(u), z(u) where $\alpha = y + z$ and $\beta = y - z$. If we assume that $\omega_v \not\equiv 0$ we may write the system (2.18) in the form

$$y'' = (\hat{a} - 1)y - 2y(y^2 - z^2)$$
$$z'' = \hat{a}z - 2z(y^2 - z^2) \tag{3.6}$$

where $\hat{a}$ is a constant. [Note: the constant $\hat{a}$ in (3.6) corresponds to a + 1/2 in (2.17)]. One sees at once that this system is reducible. Setting $y(u) \equiv 0$ leaves us with the O.D.E. for z(u)

$$z'' = \hat{a}z + 2z^3$$

while setting $z(u) \equiv 0$ gives us the O.D.E.

$$y'' = (\hat{a} - 1)y - 2y^3.$$

These special solutions produce surfaces of Joachimsthal type with planar lines of curvature in the u-direction. For example, in the case where $y(u) = 0$ we have $\omega_u = z(u)\sinh\omega$ which gives the integral identity $(\omega_u/\sinh\omega)_v = 0$. From this one concludes that there is a function $\hat{z}(v)$ with $\omega_v = \hat{z}(v)\sinh\omega$. The complete solution to this problem has been given by U. Abresch [1] and also R. Walter [11]. Abresch also looked at the case when $z(u) \equiv 0$.

The system (3.6) has two integrals, namely

$$(3.7)\qquad \begin{aligned} &y'^2 - z'^2 - (\hat{a} - 1)y^2 + \hat{a}z^2 + (y^2 - z^2)^2 = h \\ &(zy' - yz')^2 + z'^2 + z^2(y^2 - z^2 - \hat{a}) = k. \end{aligned}$$

We use the first of these integrals to introduce the Hamiltonian for the system (3.6)

$$(3.8)\qquad \mathscr{H}(y,z,p_y,p_z) = \frac{1}{4}(p_y^{\,2} - p_z^{\,2}) - \psi(y,z) = h$$

where $\psi(y,z) = (\hat{a} - 1)y^2 - \hat{a}z^2 - (y^2 - z^2)^2$.

We now solve the Hamilton-Jacobi equation. The change of variables (2.24) in the present context is given by

$$(3.9)\qquad \begin{aligned} y^2 &= -(1 - s)(1 - t) \\ z^2 &= -st. \end{aligned}$$

This transformation gives a diffeomorphism between any one of the quadrants in the y - z plane and the region defined by $s > 1$ and $t < 0$. The Hamilton - Jacobi Equation (2.23) may be written

$$(3.10)\qquad \frac{1}{4}(\theta_y^{\,2} - \theta_z^{\,2}) = (\hat{a} - 1)y^2 - \hat{a}z^2 - (y^2 - z^2)^2 + h$$

where $\Theta(y,z,h,k)$ is to be found. This is transformed into

$$(3.11)\qquad \begin{aligned} &s(s - 1)\Theta_s^{\,2} - t(t - 1)\Theta_t^{\,2} = g(s) - g(t) \\ &g(s) = -s^3 + (\hat{a} + 1)s^2 + (-\hat{a} + h)s + k \end{aligned}$$

which is just (2.25) where $\hat{a}$ in (3.11) corresponds to a + 1/2 in (2.25). The relation may be written

$$(3.12) \qquad 4g(s) = 4\,(-s + p)^3 - I(-s + p) + J$$

where $p = (\hat{a} + 1)/3$ and the separated variable equations (2.26) are written as

$$(3.13) \qquad \begin{aligned} s'(\lambda)^2 &= s(s - 1)g(s) \qquad s > 1 \\ t'(\lambda)^2 &= t(t - 1)g(t) \qquad t < 0 \end{aligned}$$

with (2.27) $\lambda'(u) = 2/(s - t)$.

A key role in the solution for $\omega(u,v)$ is played by the invariants I, J with discriminant $\Delta = I^3 - 27J^2$.

<u>Case 1</u>: $\Delta < 0$. In this case the cubic polynomial g(s) has exactly one root. If we are to generate a real solution using (3.11) this root must be greater than one, $g(s) = -(s - s_M)h(s)$ where h(s) is positive for all s. This leads to solutions $s(\lambda)$, $t(\lambda)$ where $s(\lambda)$ is periodic (with period 2L) oscillating between 1 and s_M, while $t(\lambda)$ will be periodic (with period 2M) with vertical asymptotes at $t= -\infty$.

Figure 1

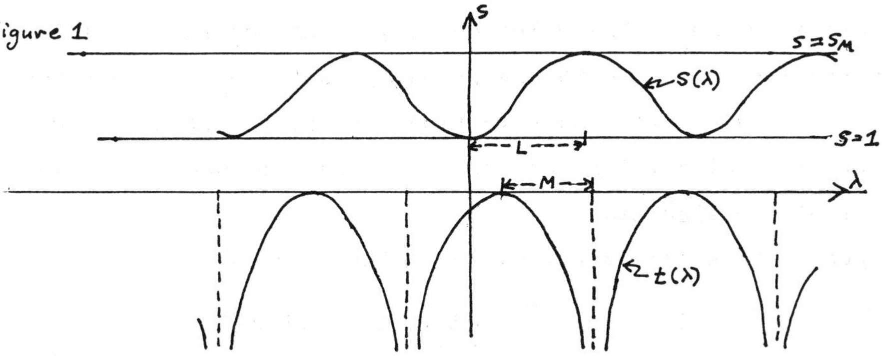

The half periods are computed using (3.13).

$$(3.14) \qquad L = \int_1^{s_M} \frac{ds}{\sqrt{s(s-1)g(s)}}, \qquad M = \int_{-\infty}^{0} \frac{dt}{\sqrt{t(t-1)g(t)}}$$

Furthermore, there is an arbitrary phase parameter between $s(\lambda)$, $t(\lambda)$ corresponding to the constant k' in $\partial\theta/\partial k = k'$.

The transformation (3.7) gives four possible choices for $\langle y,z\rangle$ given $\langle s,t\rangle$ namely $\langle y,z\rangle$, $\langle -y,-z\rangle$, $\langle y,-z\rangle$, $\langle -y,z\rangle$. Let $p_1(X)$, $p_2(X)$, $p_3(X)$, $p_4(X)$ be the corresponding polynomials obtained from (2.20). We find that $p_2(X) = p_1(-X)$. For Case 1 the discriminant Δ is negative. This means that $p_i(X)$ has exactly two real roots, both positive or both negative. We want two positive roots. Furthermore $p_4(X) = X^4 p_1(1/X)$ will have two positive roots if $p_1(X)$ does. The choice of $p_4(X)$ will lead to the solution of $-\omega(u,v)$ of (3.2) which gives the conformal factor for the dual cmc immersion $z(u,v) = x(u,v) + 2\xi(u,v)$. Thus if $\Delta < 0$ the choice of $\langle y,z\rangle$ is essentially determined.

The following lemma describes the behavior of $\omega(u,v)$ near a vertical asymptote of $t(\lambda)$.

<u>Lemma 3.2</u>: Let $t(\lambda)$ have a negative vertical asymptote at λ_o. By solving (3.13) for $\lambda = \lambda(u)$ there will be a vertical asymptote at $u = u_o$. The solution $\omega(u,v)$ extends analytically across the line $u = u_o$ with the property that $\omega(u_o,v) = \text{constant}$. The functions $y(u)$, $z(u)$ have vertical asymptotes at $u = u_o$ with the property that $z'(u)$ is negative in a neighborhood of u_o and $y'(u)$ is of one sign near u_o.

<u>Proof</u>: Equation (2.20) can be written in the form

$$2\omega_v^2 = -(1 + y^2 + z^2)\cosh 2\omega - 2yz\sinh 2\omega \tag{3.15}$$
$$-4z'\cosh\omega - 4y'\sinh\omega + 3\gamma.$$

The right side can be written as $-(1 + y^2 + z^2)Q$ where Q is a linear expression in the hyperbolic functions. The expression for Q has a definite limit as $u \to u_o$. If we define ω_o by

$$\tanh\omega_o = \sqrt{s_o - 1} / \sqrt{s_o}$$

where $s_o = s(u_o)$ then the limit for Q is

a) $\cosh 2(\omega - \omega_o) - 4\cosh(\omega - \omega_o) + 3$, if $z' < 0$, $y' > 0$

b) $\cosh 2(\omega + \omega_o) - 4\cosh(\omega + \omega_o) + 3$, if $z' < 0$, $y' < 0$

c) $\cosh 2(\omega - \omega_o) + 4\cosh(\omega - \omega_o) + 3$, if $z' > 0$, $y' < 0$

d) $\cosh 2(\omega + \omega_o) + 4\cosh(\omega + \omega_o) + 3$, if $z' > 0$, $y' > 0$

In cases (c), (d) the expression is positive and so (3.15) would have no solution near $u = u_o$. In case (a) the right side will have the single solution $\omega = \omega_o$ at $u = u_o$ so that $\omega(u_o,v) = \omega_o$ while in case (b) we find $\omega(u_o,v) = -\omega_o$

Q.E.D.

If the discriminant Δ is negative the solution $\omega(u,v)$ will be periodic in the v-direction with finite period $2B$ where we may suppose $\omega_v(u,kB) = 0$ and $\omega(u,kB + v) = \omega(u,kB-v)$. In general $\omega(u,v)$ will be quasi-periodic in the u-direction being periodic only if the ratio L/M of the periods for $s(\lambda)$, $t(\lambda)$ is a rational number. There will be an infinite sequence $\{A_m\}$ extending to infinity in both directions where $\omega(A_m, v) = \omega_m$, a constant.

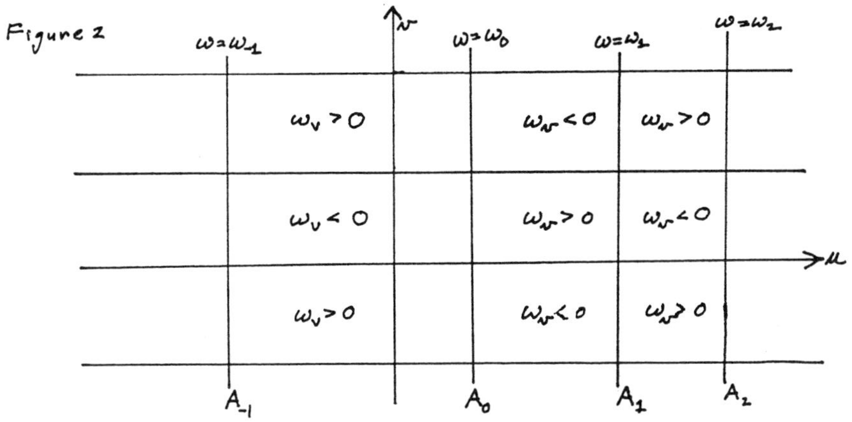

Let us consider the immersed surface itself in R^3. For Δ negative the polynomial equation (2.20) has two positive roots so that the period in the v-direction is finite. The immersion $x(u,v)$ is determined from the fundamental form (3.1).

Theorem 3.1: Suppose the solution $\omega(u,v)$ to (3.2) satisfies the symmetry condition $\omega_v(u,kB) = 0$ for all u, k any integer, and B positive. Let $x(u,v)$ be the corresponding $H = 1/2$ immersion into R^3 with fundamental form (3.1). The following are true.

a) $\omega(u,kB + v) = \omega(u,kB - v)$.

b) There are planes Ω_k such that $\mathbf{x}(u,kB) \subset \Omega_k$ and the immersion is equivariant with respect to reflections about these planes

$$x(u,kB - v) = \mathcal{R}_k \circ x(u,kB + v) \tag{3.16}$$

where $\mathcal{R}_k$ is the reflection map in R^3 about Ω_k.

c) If the immersion is of Enneper type so that $\omega(u,v)$ also satisfies (3.3), then the planes of symmetry Ω_k all contain a common line ℓ, the axis. The centers $p(u)$ of the spheres $S[p(u), R(u)]$ lie on this axial line.

Proof: Assertions (a) and (b) are immediate consequences of the Cauchy-Kowaleski theorem and Bonnet's theorem (Theorem 2.1).

Either the planes Ω_k are all parallel or there is a common fixed angle between two successive planes. In the latter case the conclusion follows. If the planes are parallel they must be identical since the curves $x(\hat{u},v)$ for any $\hat{u}$ are spherical. However, the axial line is not immediately seen.

However, one can calculate directly that the centers $p(u)$ of the spheres all lie on a given line which must be the axis. To see this one starts with equation (3.3)

$$p(u) - x(u,v) = x_o e_1 + z_o e_3$$

where $x_o = R\sin\theta = -1/z$ and $z_o = R\cos\theta = (z - y)/z$ are functions of u. Let X be the matrix $[e_1, e_2, e_3]^T$ where $\{e_i\}$ is the orthonormal Darboux frame associated with the immersion. One finds $X_u = \Omega X$ where

$$\Omega = \begin{bmatrix} 0 & -\omega_v & \sinh\omega \\ \omega_v & 0 & 0 \\ -\sinh\omega & 0 & 0 \end{bmatrix}$$

Upon differentiating we find $p_u = \varphi(u)F$ where $\varphi(u) = 1/z^2$ and $F = (A,B,C)X$ where

$$A = z' + z^2\cosh\omega + yz\sinh\omega$$

$$B = \omega_v z$$

$$C = (z'y - zy') - z\sinh\omega$$

A direct calculation shows that $(A,B,C)_u = (A,B,C)\Omega^T$ from which one concludes that $p_{uu} = \varphi'(u)F$. Furthermore $|F|^2 = k$ where k is the constant appearing the expression (3.9) for g(s). Thus p_{uu} is parallel to p_u and the centers p_u all lie on a fixed line. We have

$$p_u = (\sqrt{k}/z^2)f \tag{3.17}$$

where f is a fixed unit vector. Q.E.D.

If we introduce a coordinate system for R^3 so that f is the unit vector along the x-axis then an integration of (3.14) gives us an expression for p(u) which in turn enables us to compute the immersion x(u,v) itself using (3.3).

Now if the discriminant Δ is negative and the period L/M is rational then $\omega(u,v)$ will be doubly periodic with a rectangular period lattice, say $\omega(u + 2A, v) = \omega(u,v) = \omega(u,v + 2B)$. By Theorem 3.1 there is a paddle wheel symmetry for the immersion and we have

$$\begin{aligned} x(u + 2A,v) &= x(u,v) + \tau_1 f \\ x(u,v + 2B) &= r_{\theta_2} \circ x(u,v) \end{aligned} \tag{3.18}$$

where τ_1 is a translation parameter along the axis f and r_{θ_2} is a rotation through an angle θ_2 about the axis. Our solutions depend on 4 parameters $(\hat{a}, h, k, k')$. We have a map

$$\Phi(\hat{a},h,k,k') = (\tau_1,\theta_2). \tag{3.19}$$

One parameter is used to keep L/M rational. If we choose the remaining parameters so that $\tau_1 = 0$ and $\theta_2/2\pi$ is rational then we are left with one free parameter. We expect to find one parameter families of immersed H = 1/2 tori in R^3 which are isometrically distinct.

To be more explicit one can proceed in the following manner. Write the determining polynomial g(s) in (3.11) in the form

(3.20) $g(s) = -(s - c)[s^2 - (1 + a^2 - b^2)s + a^2] \equiv -(s - c)h(s)$.

We require that $c \geq 1$. For $c = 1$ we are in the class of Joachimsthal surfaces and in this case the parameters (a, b) are those used by Abresch in [1]. These parameters are non-negative and also satisfy $a + b \geq 1$. The discriminant Δ will be negative precisely when the parameters (a, b) lie in the infinite strip bounded by the lines $a + b = 1$, $b = a + 1$ and $b = a - 1$. If $c = 1$ the solution $\omega(u,v)$ is doubly periodic but for $c > 1$ this will be true only if the ratio M/L in (3.14) is rational. Both M and L are expressed by definite integrals which have limits as $c \to 1^+$. In fact one easily finds

$$\text{(3.21)} \quad \text{limit } (M/L) = \frac{2}{\pi}\int^{\infty} \frac{b\, du}{(1 + u^2)\sqrt{u^4 + (1 + a^2 - b^2)u^2 + a^2}} .$$

Let the fourth parameter $k' = 0$ so that $x(0,v)$ lies in a plane of reflective symmetry Π_0 perpendicular to the axis ℓ. In this case we can compute the rotation angle θ_2 using (2.20). We find

$$\text{(3.22)} \qquad \theta_2 = \int_{X_m}^{X_M} (X + X^{-1}) \frac{dX}{\sqrt{p(X)}}$$

$$p(X) = -X^4 - 4[b\sqrt{c-1} - a\sqrt{c}]X^3 + [2-4(a^2-b^2+c)]X^2 + 4[b\sqrt{c-1} + a\sqrt{c}]X - 1$$

where $X_m < X_M$ are the two positive roots of $p(X)$. If $c = 1$ Abresch derived the formula

$$\text{(3.23)} \quad \theta_2 = \frac{\pi}{2} + \int_0^{\infty} \frac{a\, du}{(1+u^2)\sqrt{u^4+(1 + a^2-b^2)u^2+b^2}}$$

Finally if M/L is rational we may compute the translation parameter τ_1. For $c > 1$ this is a complicated integration formula involving both $s(\lambda)$ and $t(\lambda)$. If $c = 1$ there is a great simplification yielding

$$(3.24)\quad \sqrt{k}\,\tau_1 = 2\sqrt{a}\int_0^{\infty} \frac{(1-2qu^2)\,du}{\left(\sqrt{u^4-2qu^2+1}\right)\left[u^2+\sqrt{u^4-2qu^2+1}\right]} = 2\sqrt{a}\,F(q)$$

where $b^2 = a^2 + 2qa + 1$. The condition $\tau_1 = 0$ reduces to $F(q) = 0$ which gives a simple quadradic condition on (a, b), a fact exploited by Abresch in [1].

The function M/L is not constant along the curve $\tau_1 = 0$ when $c = 1$ and is greater than one. We may select an integer (say $n = 2$) so that the surface M/L = 2 in (a, b, c)-space cuts the curve $\tau_1 = 0$ transversally when $c = 1$ at a point (a_o, b_o). On the surface M/L = 2 the condition $\tau_1 = 0$ is a curve C terminating in the point $(a_o, b_o, 1)$. Consider the function $\theta_2(a, b, c)$ on this curve. One needs to check that θ_2 has a non-zero rate of change along C. If so there will be an infinite number of points along C where $\theta_2/2\pi$ is rational. Now let the parameter k' vary. Changing k' removes the reflectional symmetry about the plane Π_o. Therefore we now expect each point on C for which $\tau_1 = 0$ and $\theta_2/2\pi$ is a specific rational number to evolve into a curve along which these two parameters remain constant.

Case 2: $\Delta = 0$. Here we find some beautiful examples of immersions which have cylindrical (or unduloidal) embedded ends with groups of sphere-like lobes attached in the middle. The cylindrical ended surface is pictured in the paper of Pinkall and Sterling [11]. We describe these surfaces by writing down the appropriate cubic polynomial $g(s)$ in the form (3.11). This determines the constants $\{a,h,k\}$ and also the invariants I, J. One solves for $s(\lambda)$, $t(\lambda)$ using (3.13) with $\lambda'(u) = 2/(s-t)$. A choice of initial conditions determined by the fourth constant k' allow us to solve for $y(u)$, $z(u)$ and the polynomial $p(X)$ in the formula (2.20).

Example 1: Embedded Cylindrical Ends

$$p(X) = -(X+1)^2(X-d)(X-\frac{1}{d}) \quad d > 1$$

(3.25) $$g(s) = -(s-1)(s-1+D)^2$$

$$4D = (\sqrt{d} + \frac{1}{\sqrt{d}})^2, \qquad \frac{\hat{a}+1}{3} = 1 - D.$$

Figure 3

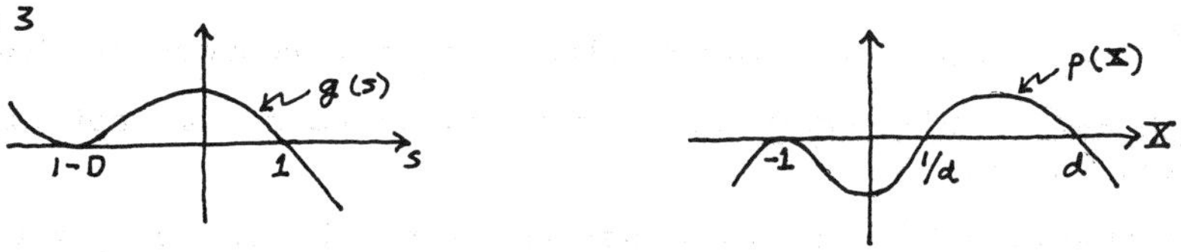

We start with g(s) and solve for s(u), t(u) with initial conditions s(0) = 1, s'(0) = 0 ($s(u) \equiv 1$ here so that the surface will be of Joachimsthal type) t(0) = 0, t'(0) = 0. This leads to y(0) = z(0) = 0 and determines the above expression for p(X) (2.20) when u = 0. The solutions for $s(\lambda)$, $t(\lambda)$ have the form

Figure 4

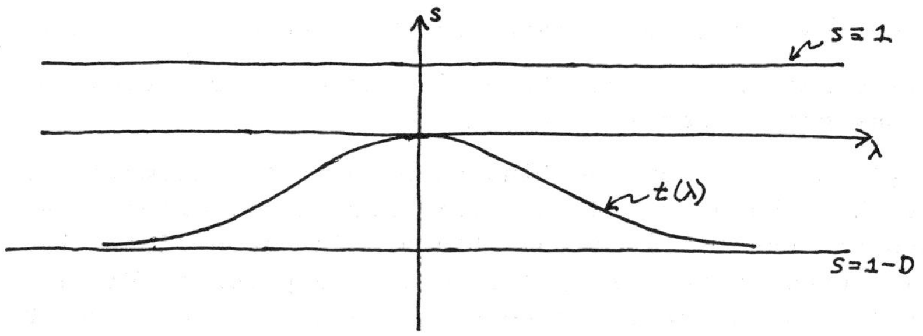

As $\lambda \longrightarrow \infty$ so does u and limit t(u) = 1 - D. We claim that $\omega(u,v) \longrightarrow 0$ uniformly as u becomes infinite. To see this one computes the limit of the polynomials p(u, X), $X = e^{\omega}$ as u becomes

infinite. In the limit we have $s = 1$, $t = 1 - D$, $s' = t' = 0$. We find using (2.20) that

$$p_{\infty}(X) = -D(X^2 - 1)^2$$

The assertion follows and the immersion becomes a covering of the round cylinder with $H = 1/2$ as u becomes infinite.

From our initial conditions we see that $\omega(u,v)$ is an even function of u with $\omega_u(0,v) = 0$. Therefore $x(0,v)$ lies in a plane of reflective symmetry Π_o for the immersion. This plane is perpendicular to all of the paddle wheels Ω_k and also the axial line ℓ.

Using the equation $4X_v^{\ 2} = p(u,X)$ when $u = 0$ we can compute the half period in the v-direction of the function $\omega(u,v)$

$$\text{(3.26)} \qquad B = \int_{1/d}^{d} \frac{2dx}{(x+1)\sqrt{(x - \frac{1}{d})(x - d)}} = \pi/D.$$

We have $\omega_v(u, kB) = 0$, $\omega_u(0,v) = 0$ and limit $\omega(u,v) = 0$ as $|u| \to \infty$.

The curves $x(u,kB)$ lie in the planes of symmetry Ω_k. We wish to measure the angle between Ω_o and Ω_1. We can measure this angle in two ways. First calculate the angle using the curve $x(0,v)$ $0 \le v \le B$ connecting Ω_o to Ω_1 and lying in the plane Π_o. For this curve the curvature is $k_2 = e^{-\omega}\cosh\omega$ and $ds = e^{\omega}dv$.

$$\text{(3.27)} \qquad \theta_o = \int_0^B \cosh\omega \, dv = \int_{1/d}^{d} \left(\frac{x + x^{-1}}{2}\right)\left(\frac{2dx}{\sqrt{p(x)}}\right)$$

As u becomes infinite, $\cosh\omega$ approaches one and we find

$$\Theta_1 = B = \int_0^B dv = \int_{1/d}^{d} \frac{2dx}{\sqrt{p(x)}} = \pi/D. \tag{3.28}$$

A direct calculation shows that $\Theta_0 + \Theta_1 = 2\pi$. The state of affairs is indicated by the following sketch.

Figure 5

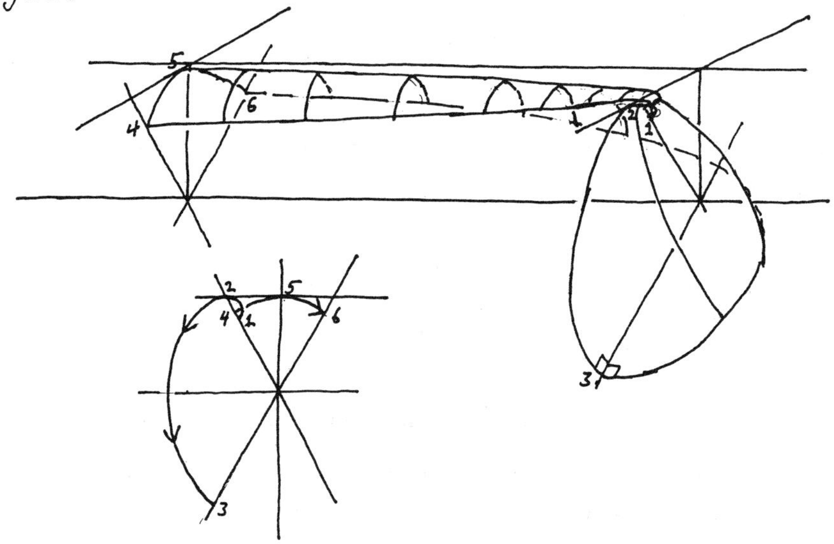

We have $\Theta_1 = \pi/D$ where $4D = \left(\sqrt{d} + \frac{1}{\sqrt{d}}\right)$ with $1 < D < \infty$. Thus for $\Theta_1 = \pi/n$ our immersed surface will have embedded cylindrical ends with n sphere-like lobes attached in the middle. In particular we may choose n = 2 and obtain an immersion with just two lobes. Since these are Joachimsthal type surfaces, explicit formulas for $\omega(u,v)$ and $x(u,v)$ can be written down in terms of elliptic functions using the methods of U. Abresch or R. Walter.

If we introduce the Abresch parameters (a,b) into the formula for g(s) in (3.20) we find that in Example 1 $b = a + 1$ and $g(s) = -(s-1)(s+a)^2$. In this case the rotation angle θ_2 and the position vector itself can be expressed in terms of elementary functions. The corresponding K = +1 surface $y(u,v) = x(u,v) + \xi(u,v)$ was first discovered by H. Sievert in 1886. A nice discussion of this and other interesting examples may be found in the book [9]. This surface has the same number of lobes as the H = 1/2 surface but the asymptotic cylinders become narrow tubes converging onto the axial line ℓ. The principal curvatures are $\hat{k}_1 = \sinh\omega/\cosh\omega$, $\hat{k}_2 = \cosh\omega/\sinh\omega$ so that the image of any coordinate line v = kB along which $\omega(u,kB) = 0$ will be a cuspoidal curve on the surface.

Example 2: Embedded Unduloidal Ends.

We now set

$$p(X) = -(X+p)^2\left(X - \frac{1}{pd}\right)\left(X - \frac{d}{p}\right)$$

(3.29) $$g(s) = -(s-1-E)(s-1+D)^2$$

$$4D = \left(\sqrt{d} + \frac{1}{\sqrt{d}}\right)^2, \quad 4E = \left(p - \frac{1}{p}\right)^2.$$

Figure 6

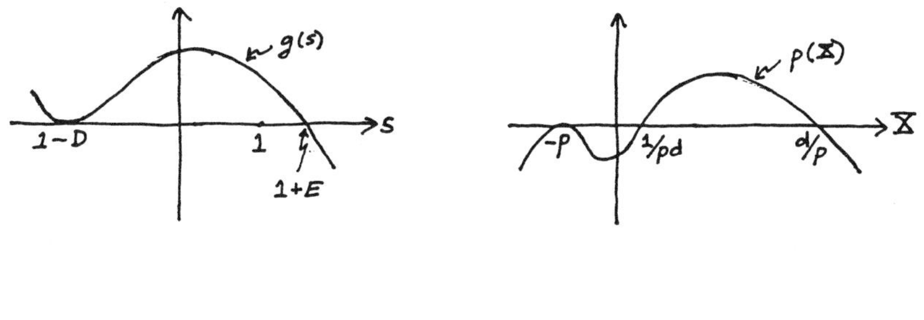

As in Example 1 we have found $p(X) = p(0,X)$ by solving for $s(\lambda)$, $t(\lambda)$ with initial conditions $s(0) = 1$, $s'(0) = 0$, $t(0) = 0$, $t'(0) = 0$.

Figure 7

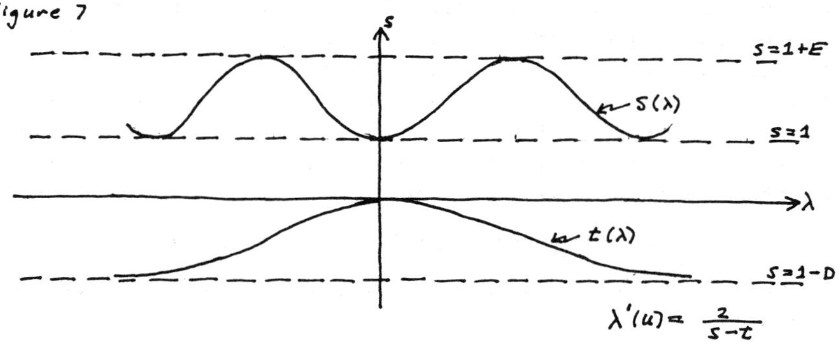

As u becomes infinite $t(u)$ approaches $1 - D$ and there is a half period B such that $\omega_v(u,kB) = 0$. For our initial conditions $\omega_u(0,v) = 0$ which again implies that $x(0,v)$ lies in a plane of reflective symmetry Π_o. At the ends $s(\lambda)$ remains oscillatory while $t(\lambda) \longrightarrow 1 - D$. This limit determines an asymptotic solution $\hat{\omega}(u,v)$. If we calculate the polynomial $p(X)$ corresponding to the values $s = 1$ and $t = 1 - D$ we find

$$(3.30) \qquad p(X) = -D(X - m)^2\left(X + \frac{1}{m}\right)^2, \quad m = \frac{\sqrt{D + E} - \sqrt{E}}{\sqrt{D}}.$$

This implies that $\hat{\omega}(u,v)$ depends only on u. The ends are unduloidal. From the equation $\hat{\omega}_{uu} + \sinh\hat{\omega}\cosh\hat{\omega} = 0$ we have

$$(3.31) \qquad \hat{\omega}_u^{\,2} + \frac{1}{2}\cosh 2\hat{\omega} = \frac{1}{2}\cosh(2\hat{\omega}_M)$$

where $\hat{\omega}_M = \hat{\omega}_{max}$. A direct calculation shows that $\cosh^2\hat{\omega}_M = 1 + E$. This determines the type of unduloid.

Once again there are two ways to compute the angle between

the planes Ω_o and Ω_1. As in the previous example $\Theta_o + \Theta_1 = 2\pi$ and

$$(3.32) \qquad \theta_1 = \frac{(1 + p^2)\pi}{\sqrt{p^4 + (d + d^{-1})p^2 + 1}} .$$

For fixed $p > 0$, Θ_1 in a decreasing function of d with $\Theta_1(1) = \pi$ and $\Theta_1(\infty) = 0$. We may choose $\Theta_1 = \pi/n$ which determines an immersed surface with embedded unduloidal Delaunay ends and n-lobes in the middle.

From our initial conditions $\omega_u(0,v) = 0$ so that $x(0,v)$ lies in a plane of reflective symmetry Π_o perpendicular to the axis. We can change the initial conditions so that $t(0) = 0$, $s(0) = s_o$. This gives $z(0) = 0$ and $y^2(0) = s_o - 1$. Going back to (3.5) we find that $R(0)$ is infinite and $\tan \Theta = 1/y = 1/\sqrt{s_o - 1}$. Thus $x(0,v)$ still lies in a plane perpendicular to the axis but now the immersion cuts through the plane at a fixed angle Θ_o where $\tan\Theta_o = 1/\sqrt{s_o - 1}$.

For each integer n there is a one-parameter family of Enneper type $H = 1/2$ immersions with embedded unduloidal Delaunay ends of assigned type with n sphere-like lobes in the middle.

Finally, we observe that for each of the Enneper type cmc surfaces constructed in this section there is an associated surface of the same type obtained by switching the second fundamental form. We now set

$$(3.33) \qquad \begin{aligned} ds^2 &= e^{2\omega}(du^2 + dv^2) \\ II &= e^{\omega}(\cosh\omega\, du^2 + \sinh\omega\, dv^2) \end{aligned}$$

so that $k_1 = e^{-\omega}\cosh\omega$ and $k_2 = e^{-\omega}\sinh\omega$. Again we suppose that $x(\hat{u},v) \subset S[p(\hat{u}), R(\hat{u})]$ and set as before

(3.3) $$p(u) - x(u,v) = x_o(u)\, e_1 + z_o(u) e_3$$

An application of Lemma 3.1 yields

(3.34) $$\omega_u = y(u)\cosh\omega + z(u)\sinh\omega$$

where $y = -1/x_o$ and $z = (z_o - 1)/x_o$. Here $x_o = R\sin\theta$ and $z_o = R\cos\theta$, giving us

(3.35) $$R^2 = \frac{1+(z-y)^2}{y^2}, \quad \tan\theta = \frac{-1}{y-z}.$$

For the corresponding K = +1 surface

(3.36) $$\hat{R}^2 = \frac{1 + z^2}{y^2}, \quad \tan\hat{\theta} = 1/z.$$

The claims of theorem 3.1 are still valid. The calculation showing that the centers **p**(u) all lie on an axial line is entirely similar. One now finds that $\mathbf{p}_u = (1/y^2)\mathbf{F}$ where **F** is a constant vector. In terms of the Darboux frame we now have

$\mathbf{F} = A\mathbf{e}_1 + B\mathbf{e}_2 + C\mathbf{e}_3$

(3.37)
$$\begin{aligned} A &= y' + yz\cosh\omega + y^2\sinh\omega \\ B &= y\omega_v \\ C &= (y'z - yz') - y\cosh\omega \end{aligned}$$

We find that $|\mathbf{F}|^2 = h + k$ so that

$$\mathbf{p}_u = (\sqrt{h+k}/y^2)\mathbf{f}$$

where **f** is a constant unit vector.

IV. MINIMAL SURFACES IN R^3

Let $x(u,v)$ be a conformally immersed minimal surface in R^3 parameterized by its lines of curvature and with fundamental forms

(4.1)
$$ds^2 = e^{2\omega}(du^2 + dv^2)$$
$$II = -du^2 + dv^2$$

and principal curvatures $k_1 = -e^{-2\omega}$, $k_2 = e^{-2\omega}$ where $\omega(u,v)$ is a solution to the Gauss Equation

(4.2)
$$\Delta\omega - e^{-2\omega} = 0.$$

We also suppose that our immersion is of Enneper type so that the auxillary equation (2.17) is satisfied.

(4.3)
$$2\omega_u = \alpha(u)e^{\omega} + \beta(u)e^{-\omega}$$

where $\alpha(u)$, $\beta(u)$ are solutions to the system (2.18) where $A = 0$ and $B = 1$. We seek to represent these minimal surfaces using the Weierstrass representation.

If $x(w)$, $w = u + iv$ is our minimal surface in conformal parameters then $x(w)$ is a harmonic vector function. Let $x^*(w)$ be its harmonic conjugate and set $\hat{x}(w) = x(w) + ix^*(w)$. We write

$$\hat{x}'(w) = x_u + ix^*_u = x_u - ix_v = 2x_w$$

and the conformality condition $(\hat{x}'(w),\hat{x}'(w)) = 0$ where we use the non-Hermetian inner product. The Weierstrass representation asserts that there is a pair of meromorphic functions $f(w)$, $g(w)$ such that

(4.4)
$$\hat{x}'(w) = \Phi(w) = \langle \frac{f}{2}(1 - g^2),\ \frac{if}{2}(1 + g^2),\ fg \rangle .$$

With this representation we have the following identites

(4.5) $$ds^2 = \frac{|f|^2}{4}(1 + |g|^2)^2|dw|^2$$

$$K = \frac{-16\,|g'|^2}{|f|^2(1 + |g|^2)^4}.$$

<u>Lemma 4.1</u>: If x(u,v) is a minimal surface conformally parameterized by its lines of curvature and satisfying (4.1-2) and if its Weierstrass representation is as above then

(4.6) $$fg' \equiv 1.$$

<u>Proof</u>: The Hopf function (2.2) associated with the second fundamental form is

(4.7) $$\varphi(w) = \left(\frac{L - N}{2}\right) - iM = -2(x_w \cdot \xi_w) = -1.$$

However, we have $2x_w = \hat{x}'(w) = \Phi(w)$ from (4.4) and we also have

(4.8) $$\xi(w) = \left(\frac{2\mathrm{Re}(g)}{1 + |g|^2}, \frac{2\,\mathrm{Im}(g)}{1 + |g|^2}, \frac{|g|^2 - 1}{|g|^2 + 1}\right).$$

A direct calculation gives $2(x_w \cdot \xi_w) = fg'$.

Q.E.D.

It is convenient to restate the Weierstrass representation by setting F = fg. We have

(4.9) $$\Phi(w) = \left(\frac{F}{2}(g^{-1} - g), \frac{iF}{2}(g^{-1} + g), F\right)$$

where g(w) is to be recovered from F(w) by solving the differential equation

(4.10) $$F = fg = g/g' \text{ or } g' = \frac{1}{F}g.$$

We have the following result.

Theorem 4.1. Let $x(u,v)$ be a minimal surface of Enneper type whose representation satisfies the equations (4.1-3). Suppose that the axial line ℓ is chosen to be the z-axis with Weierstrass representation in the form (4.9). There is a non-zero real number c such that $F(w) = cG(w/c)$ where $G(w)$ is a solution to the differential equation

$$G'(w)^2 = -G^3 + pG^2 + qG + 1 \tag{4.11}$$

where p,q are real constants.

Note: The constant c is just a homothety and rescaling parameter and so without loss of generality we may assume $c = 1$.

Proof: The functions $\alpha(u)$, $\beta(u)$ are solutions of the system (2.18) where $A = 0$, $B = 1$, and if $X = e^{\omega}$ we may write (2.19) as $4X_v^{\,2} = p(u,X)$ where

$$p(u,X) = -\alpha^2X^4 - 4\alpha'X^3 + 6\gamma X^2 + 4\beta'X - (4 + \beta^2).$$

Suppose $x(u,v) \subset S[\mathbf{p}(u), R(u)]$ the sphere with center $\mathbf{p}(u)$ and radius $R(u)$. As in Section III we may apply Joachimsthal's Lemma 3.1 and (3.3-4). We have

$$\mathbf{p}(u) - x(u,v) = x_o(u)\mathbf{e}_1 + z_o(u)\mathbf{e}_3$$

where $\{\mathbf{e}_1,\mathbf{e}_2,\mathbf{e}_3\}$ is the Darboux frame associated with the mapping and $x_o^{\,2} + z_o^{\,2} = R^2$, $\tan\theta = x_o/z_o$. We find that $x_o = -2/\alpha$, $z_o = -\beta/\alpha$ so that

$$R^2 = \frac{4 + \beta^2}{\alpha^2} \qquad \tan\theta = 2/\beta. \tag{4.12}$$

As in Section III, Theorem 3.1, the vector $\mathbf{f} = \mathbf{p}_u/||\mathbf{p}_u|$ is a constant unit vector parallel to the axis ℓ. Select a coordinate system so that f is parallel to the z-axis. This means that $z(u,v) = x(u,v)\cdot f$ and $F(w) = 2(z_u - iz_v)$. We now show that

$$(4.13) \qquad 2\sqrt{k}z_u = e^{\omega}(2\alpha' + \alpha^2 e^{\omega} - \alpha\beta e^{-\omega})$$

$$2\sqrt{k}z_v = 2\alpha\omega_v e^{\omega} = (2\alpha e^{\omega})v$$

whence

$$2\sqrt{k}z(u,v) = 2\alpha e^{\omega} + \int 2\alpha\beta du.$$

One derives (4.13) by first computing p_u using (3.3) keeping in mind that $X_u = \Omega X$ where $X = [e_1, e_2, e_3]^T$ and

$$\Omega = \begin{bmatrix} 0 & -\omega_v & -e^{-\omega} \\ \omega_v & 0 & 0 \\ e^{-\omega} & 0 & 0 \end{bmatrix}$$

We find

$$(4.14) \qquad p_u = (1/\alpha^2)[Ae_1 + Be_2 + Ce_3] = (1/\alpha^2)F$$

$$A = 2\alpha' + \alpha^2 e^{\omega} - \alpha\beta e^{-\omega}$$

$$B = 2\alpha\omega_v$$

$$C = 2\alpha e^{-\omega} + (\alpha'\beta - \alpha\beta').$$

As in Section III one checks that p_{uu} is parallel to p_u and so p_u is parallel to the axis ℓ. But

$$|F|^2 = (\alpha'\beta - \alpha\beta')^2 + 4\alpha'^2 + 4\alpha^2(\alpha\beta - a) = 4k$$

where k is constant. It is an integral to the system (2.18) when A = 0, B = 1 and is the constant term in the cubic polynomial g(s) in formula (2.25). One now calculates $x_u \cdot f$ and $x_v \cdot f$ where $x_u = e^{\omega}e_1$ giving us (4.13).

Solutions to the system (2.18) are determined by the initial values. We note that $\alpha(0) = \alpha'(0) = 0$ gives us $\alpha(u) \equiv 0$, $\beta'' = \alpha\beta$ [the Hamiltonian system with $A = 0$, $B = 1$ is reducible]. In this case $R(u) = +\infty$ and the lines of curvature are all planar. We obtain the minimal surfaces of Joachimsthal type. These are the well-known Bonnet minimal surfaces [7].

Now choose parameters so that $\alpha(0) = 0$, $\alpha'(0) \neq 0$. We show later, Lemma 4.2, that this may always be done. Using (4.13) we find that $z_v(0,v) = 0$ which means that $F(iv) = 2z_u(0,v)$ is real. Now $g(w) = F(w)g'(w)$ so

$$g(w) = g_o \exp\left\{ \int_{v_o}^{v} [1/F(\xi)]d\xi \right\} \tag{4.15}$$

In particular we find

$$g(iv_2)/g(iv_1) = \exp\left\{ i\int_{v_1}^{v_2} [1/F(it)]dt \right\}$$

and thus $|g(iv)| = M$ a constant.

Since $\alpha(0) = 0$ we have

$$4X_v^{\,2} = -4\alpha' X^3 + 6\gamma X^2 + 4\beta X - (4 + \beta^2) \tag{4.16}$$

when $u = 0$. From (4.5) we see

$$2e^{\omega} = 2X = |F|(|g| + |g|^{-1})$$

$$X(0,v) = \left(\frac{M + M^{-1}}{2}\right)|F(iv)| \equiv \lambda|F(iv)|.$$

But $F(iv)$ is real so that $X(0,v) = \pm \lambda F(iv)$ where $\lambda \geq 1$. This gives from (4.16)

$$-4\lambda^2 F'(w)^2 = (-4\alpha')(\lambda^3 F^3) + (6\gamma)(\lambda^2 F^2) + (4\beta')(\lambda F) - (4 + \beta^2)$$

when $u = 0$. We now check that $4 + \beta^2 = 4\lambda^2$ when $u = 0$. To see this we observe from (4.12) that $x(0,v)$ is a planar curve and that the surface intersects the plane at an angle θ where $\tan\theta = 2/\beta$ We have selected the axis ℓ to be the z-axis so that this plane is horizontal. From the Weierstrass representation (4.8) we compute $\cos\theta = (|g|^2 - 1)/(g^2 + 1) = (M^2 - 1)/(M^2 + 1)$. Therefore we get $4 + \beta^2 = (M + M^{-1})^2 = 4\lambda^2$. Now replace $F(w) = c\, G(aW)$ where c is chosen so that $\lambda\alpha'(0)c^3 = -1$ and a is chosen so that $ac = 1$. We obtain

$$G'(w)^2 = -G^3 + pG^2 + qG + 1$$

where p,q are real constants. We have proven this identity along the imaginary axis. By analyticity it is true everywhere.

Q.E.D.

<u>Lemma 4.2</u>: Let $\langle\alpha((u), \beta(u)\rangle$ be solutions to the system (2.18) with $A = 0$, $B = 1$

$$\alpha'' = a\alpha - 2\alpha^2\beta$$

$$\beta'' = a\beta - 2\alpha\beta^2 - 2\alpha$$

from which an immersed minimal surface of Enneper type is produced. There is a value u_o for which $\alpha(u_o) = 0$.

<u>Proof</u>: We show this by solving the system using the method of separation of variables as discussed in Section II. The transformation (2.24) in this case is

$$\alpha\beta = s + t$$

$$\alpha^2 = -st$$

This map gives a diffeomorphism of the half-plane $\alpha > 0$ onto the quadrant $s > 0$, $t < 0$. Note that if $\langle\alpha,\beta\rangle$ is a solution then so is $\langle-\alpha, -\beta\rangle$. Set $g(s) = -s^3 + as^2 + hs + k$ (see 2.25) and solve the O.D.E.'s (2.26).

$$s'(\lambda)^2 = s^2 g(s), \quad s > 0$$

$$t'(\lambda)^2 = t^2 g(t), \quad t < 0 \qquad \lambda'(u) = \frac{2}{s - t}$$

Suppose, for example, that g(s) has one real root which is positive. In this case s(λ) and t(λ) will have aympototic limits equal to zero as λ becomes infinite. From the differential equation $\lambda'(u) = 2/(s - t)$ one easily verifies that u has a finite limit u_o as λ becomes infinite. This means that $\alpha(u_o) = 0$.

Other cases are treated in similar fashion. Suppose that g(s) has three real roots $\hat{a} < \hat{b} < 0 < \hat{c}$.

Figure 8

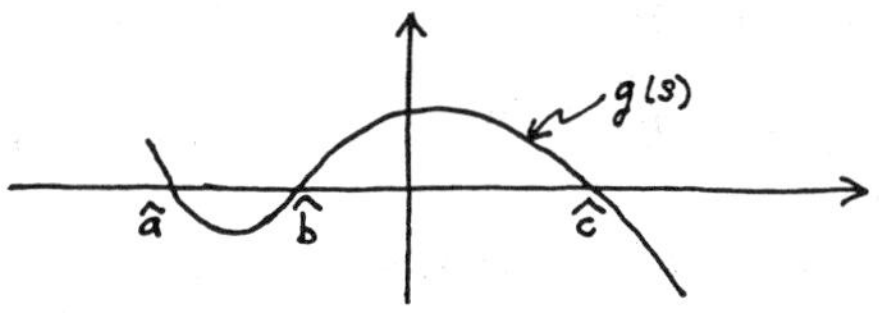

The solution for s(λ) is well determined up to a translation parameter, but there are two apparent choices for t(λ). If t(λ) has a minimum equal to $\hat{b}$ then it will be asymptotically zero at infinity and the previous discussion applies. Otherwise t(λ) will have a maximum of $\hat{a}$. Suppose $t(0) = \hat{a}$ and $s(0) = \hat{c}$. In this case the polynomial p(0,X) in (2.19) will be negative for all X implying that there is no solution for ω(u,v). In fact $\alpha\beta = s + t = \hat{a} + \hat{c}$, $\alpha^2 = - st = - \hat{a}\hat{c}$ and $\alpha'(0) = \beta'(0) = 0$. We also calculate $6\gamma = 2\hat{a} - 4\hat{b} + 2\hat{c}$ and end up with

$$p(X) = \hat{a}\hat{c}X^4 + [2(\hat{a} + \hat{c}) - 4\hat{b}]X^2 + \frac{(\hat{a} - \hat{c})^2}{\hat{a}\hat{c}} .$$

from which we see that $p(X)$ is negative for all X. For this case $g(s)$ has 3 distinct real roots and therefore all polynomials $p(u,x)$ will have either 4 real roots or no real roots at all. If we change the initial conditions for $s(\lambda)$ so that $s'(0) \neq 0$ it follows by continuity that the polynomial $p(0,X)$ will continue to be negative.

The remaining case follows by a similar argument.

Q.E.D.

<u>Theorem</u> <u>4.2</u>: Let $F(w)$ be a solution to

$$F'(w)^2 = - F^3 + pF^2 + qF + 1 \tag{4.11}$$

where p,q are real. Then $F(w) = b - 4P(w)$ where $P(w)$is the Weierstrass $\wp$-function with $P'(w)^2 = 4P^3 - IP - J$. Here

$$b = -p/3, \; I = \frac{p^2 + 3q}{12} , \; b^3 - 4bI - 16J = 1. \tag{4.17}$$

Conversely, let $P(w)$ be the Weierstrass $\wp$-function where I, J are real and let b (real) be chosen to satisfy the cubic equation above. Then $F(w)$ is a solution to (4.11) with p,q determined by the remaining two equations.

<u>Proof</u>: Suppose $F(w)$ solves (4.11) and set $4P(w) = b - F(w)$. If we set $b = p/3$ then $P(w)$ is a solution to $P'(w)^2 = 4P^3 - IP - J$ where $12I = 3q + p^2$ and $b^3 - 4bI - 16J = 1$. $P(w)$ is, up to a translation, the standard Weierstrass $\wp$-function. Conversely, given $P(w)$ with I,J real one finds b(real) satisfying solution to (4.11). Q.E.D.

Having found that $F(w) = b - 4P(w)$ we now use this formula in the Weierstrass representation (4.9). We obtain the following result.

<u>Theorem</u> <u>4.3</u>: Let $F(w) = b - 4P(w)$ be a solution to (4.11). The Weierstrass formula (4.9-10) is a triple of single-valued meromorphic functions on the complex plane analytic except at the

poles of $F(w)$. The residues of $\Phi(w)$ are all zero and consequently $\hat{x}(w)$ is a triple of meromorphic functions analytic at the poles of $F(w)$. At each pole of $F(w)$ the minimal surface $x(w)=\mathcal{R}e[\hat{x}(w)]$ has a flat end whose asymptotic plane can have any inclination except being horizontal. It cannot be perpendicular to the z-axis. Finally, $\Phi(w)$ is never equal to zero and thus $x(w)$ is a complete immersed minimal surface without branch points.

<u>Proof</u>: First we observe that $F(w) = b - 4P(w)$ is a meromorphic function with poles of order two and no residues. From the Weierstrass formula (4.9) we conclude that $\hat{z}(w)$ is a single-valued meromorphic function on the complex plane. Since $F(w)$ is a solution to (4.11) we see that $F'(w)^2 = 1$ whenever $F(w) = 0$. $F(w)$ has at most simple zeros. We may write

$$g(w) = M\exp\left\{ \int_0^w [1/F(t)]dt \right\} .$$

The residue $(1/F)$ about any zero of F is ± 1 and so $g(w)$ is a single-valued meromorphic function. Since $F(iv)$ is real we have that $|g(iv)| = M$. We also note that we have a three-parameter family of distinct immersions with parameters I, J, and M. The fourth parameter corresponds to the homothety and rescaling previously discussed.

Since the zeros of $F(w)$ are all simple, we can show that at a zero of $F(w)$, $g(w)$ will have a simple zero (or pole). This implies that the functions $F(g^{-1} - g)$ and $iF(g^{-1} + g)$ are analytic and non-vanishing across any zero of $F(w)$.

Consider what happens at a pole of F which we may assume is $w = 0$. Since $P(w)$ has a pole of order two without residue at $w = 0$ we write

$$P(w) = \frac{1}{w^2} + \sum a_n w^{2n}$$

From the formula above we find

$$g(w) = M(1 + c_3 w^3 + \ldots), \quad M > 0.$$

This leads to

$$\Phi(w) = \langle \frac{2(M - M^{-1})}{w^2}, \frac{-2i(M + M^{-1})}{w^2}, \frac{-4}{w^2} \rangle + \Phi_a$$

where Φ_a is analytic about $w = 0$. Therefore, $\Phi(w)$ has a residue of zero at all poles and $\hat{x}(w)$ is a single-valued meromorphic function.

Now $g(0) = M$ and so there is a limit unit normal vector

$$\xi(0) = \langle \frac{2M}{1 + M^2}, 0, \frac{M^2 - 1}{M^2 + 1} \rangle \equiv \xi_o$$

using (4.8). From this we find that $\hat{x}(w) \cdot \xi_o$ is analytic in a neighborhood of $w = 0$. This means that the end at $w = 0$ is flat with ξ_o being the unit normal vector to the aymptotic plane.

Q.E.D.

Example 1: The basic example is obtained by setting I and J to be zero so that the Weierstrass $\mathcal{P}$-function satisfies $P'^2 = 4P^3$ or $P(w) = 1/w^2$. From (4.12) we find $b = 1$ so that

$$F(w) = 1 - \frac{4}{w^2}. \tag{4.18}$$

Also the equation $g = Fg'$ leads to

$$g(w) = Me^w \left(\frac{2 - w}{2 + w}\right) \tag{4.19}$$

where (without loss of generality) we may assume that M is a positive constant.

At $w = 0$ the immersion has an asymptotically flat end, Λ_o, whose unit normal vector is

$$(4.20) \qquad \xi_o = \left\langle \frac{2M}{1 + M^2},\ 0,\ \frac{M^2 - 1}{M^2 + 1} \right\rangle.$$

Also the curve $\mathbf{x}(0,v)$ will lie in a horizontal plane Π_o and the surface cuts this plane at an angle Θ_o, $\tan\Theta_o = 2M/(M^2 - 1)$.

In the symmetric case when $M = 1$, we have $\xi_o = \langle 1, 0, 0\rangle$ so that Λ_o is a vertical plane while the horizontal plane Π_o is a plane of reflective symmetry. We find

$$(4.21) \qquad \hat{x}(w) = \left\langle \cosh w - \frac{4\sinh w}{w},\ -i\sinh w + \frac{4i\cosh w}{w},\ w + \frac{4}{w} \right\rangle$$

where $x(w) = \mathcal{Re}[\hat{x}(w)]$. If $|w| >> 1$ we see that $\hat{x}(w) \cong \langle \cosh w, -i\sinh w, w\rangle$ so that $x(w) \cong \langle \cosh u \cos v, \cosh u \sin v, u\rangle$.

In a neighborhood of $w = 0$ we find

$$\hat{x}(w) = \left\langle -3 - \frac{w^2}{6} + \text{--},\ \frac{4i}{w} + iw + \text{--},\ \frac{4}{w} + w \right\rangle$$

$$x(u,v) \cong \left\langle -3 - \frac{(u^2 - v^2)}{6} + \text{--},\ \frac{-4v}{u^2 + v^2} - v + \text{--},\ \frac{4u}{u^2 + v^2} + u \right\rangle.$$

The asymptotic plane is $x = -3$ and the plane of symmetry is the plane $z = 0$.

For large w the surface appears catenoidal although the surface does not close up. Each curve $x(\hat{u},v)$ for fixed $\hat{u}$ lies on a specific sphere with center on the z-axis. Near $w = 0$ the catenoidal surface is connected to the flat end $x = -3$. (See Figure 9 and attached figures at end.)

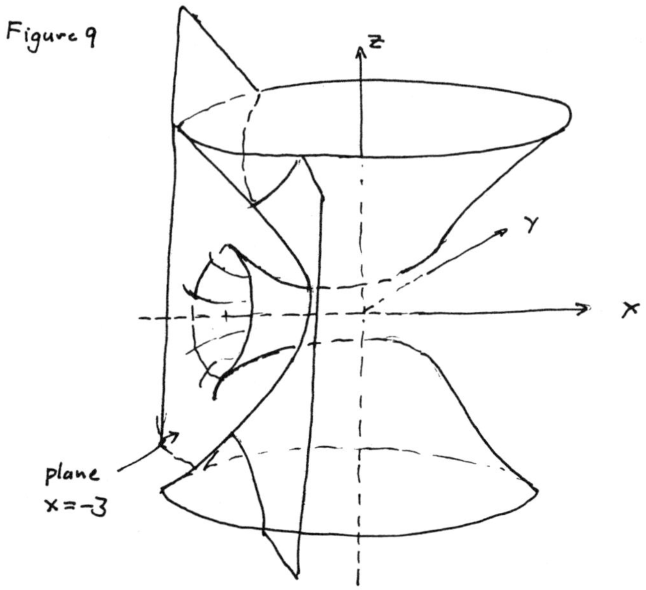

For general M the integrated equation (4.16) becomes

$$(4.22) \qquad \hat{x}(w) = \frac{1}{2}\left(\frac{e^{-w}}{M} + Me^{w}\right) + \frac{2}{w}\left(\frac{e^{-w}}{M} - Me^{w}\right)$$

$$\hat{y}(w) = \frac{i}{2}\left(\frac{e^{-w}}{M} - Me^{w}\right) + \frac{2i}{w}\left(\frac{e^{-w}}{M} + Me^{w}\right)$$

$$\hat{z}(w) = w + \frac{4}{w}\ .$$

The normal vector ξ_o at $w = 0$ is given by (4.20) and one checks that the equation of the asymtotic plane is

$$(4.23) \qquad x \cdot \xi_o = \left(\frac{2M}{1 + M^2}\right)x + \left(\frac{M^2 - 1}{M^2 + 1}\right)z = -3.$$

<u>Example 2</u>: Set $I = \frac{4r^4}{3}$, $J = \frac{-8r^6}{27}$. We find $P(w) = r^2\left[\frac{1}{\sinh^2 rw} + \frac{1}{3}\right]$ is a degenerate Weierstrass $\wp$-function with

$\Delta = I^3 - 27J^2 = 0$ and basic period $w_1 = \pi i/r$. Select B to solve $B^3 - 4IB - 16J = 1$ and set $F(w) = B - 4P(w)$. After some rearranging one can conveniently write F(w) in the form

$$(4.24) \qquad F(w) = b - \frac{c^2}{\sinh^2(cw/2)} = b - \frac{(a-b)}{\sinh^2(cw/2)}.$$

Here $a > b > 0$, $ab^2 = 1$ and $c^2 = a - b$, so that F(w) is periodic with period $w_1 = 2\pi i/c$.

The computation of g(w) gives

$$(4.25) \qquad g(w) = Me^{w/b}\left[\frac{c - \sqrt{a}\tanh(cw/2)}{c + \sqrt{a}\tanh(cw/2)}\right].$$

Suppose F(w) and g(w) are both periodic with period $w_1 = iv_o$. By integrating the Weierstrass form we find $\hat{x}(w + iv_o) = \hat{x}(w) + k$. But $x(u,v) = \mathcal{R}e[\hat{x}(w)]$ is an Enneper type surface. It follows that k is imaginary and so the real surface x(u,v) is periodic with period $(0,v_o)$.

In our example F(w) has basic period $w_1 = 2\pi i/c$. g(w) is quasiperiodic with two basic periods $ip_1 = 2\pi bi$ and $ip_2 = 2\pi i/c$. Observe that $0 < b^2c^2 = b^2(a-b) = 1 - b^3 < 1$. This shows that p_1 is less than p_2. The function g(w) will be periodic if there are relatively prime positive integers m,n such that $2\pi nb = 2\pi m/c = \vartheta$. In this case g(w) will have basic period $i\vartheta$. This condition can be written as $m/n = bc$ or $1-b^3 = (m/n)^2$ which can be solved given any integer m less than n.

In particular consider the case $m = 1$, $n > 1$. Here $\vartheta = 2\pi/c = 2\pi nb$ and F(w), g(w) both have the same basic period $i\vartheta = 2\pi i/c$. The integrated minimal surface x(u,v) will also have this period, $x\left(u, v + \frac{2\pi}{c}\right) = x(u,v)$.

We have an immersed cylinder with one flat end at $w = 0$ (mod $\frac{2\pi i}{c}$). If $w = u + iv$ and $|u| >> 1$ we find

$$g(w) \cong Me^{w/b}$$

So $g(w)$ almost has period $2\pi bi = 2\pi i/nc$. In other words for $|u| >> 1$ the immersion appears catenoidal but must wrap around the catenoid n times before closing up. The ends are not embedded.

Example 3: Consider the degenerate Weierstrass $\mathcal{P}$-function with basic period $w_1 = \pi/r$. Here $I = \frac{4r^4}{3}$, $J = \frac{8r^6}{27}$, $P(w) = r^2\left[\frac{1}{\sin^2 rw} - \frac{1}{3}\right]$. As in example 2 one ends up with the following expressions for $F(w)$, $g(w)$.

$$\begin{aligned} F(w) &= b - \frac{(b-a)}{\sin^2(cw/2)} \qquad ab^2 = 1,\ b > a,\ c^2 = b - a \\ g(w) &= Me^{w/b}\left[\frac{c - \sqrt{a}\tan(cw/2)}{c + \sqrt{a}\tan(cw/2)}\right]. \end{aligned} \tag{4.26}$$

In general we start with the Weierstrass $\mathcal{P}$-function $P(w)$ satisfying $P'(w)^2 = 4P^3 - IP - J$ where I, J are real. There are two cases depending on the sign of the discriminant $\Delta = I^3 - 27J^2$.

If w_1 and w_2 are the fundamental periods so that $P(w)$ will have poles at all periods $w = mw_1 + nw_2$ then $I = 60G_2$, $J = 140G_3$ where $G_k = \sum 1/(mw_1 + nw_2)^k$ summed over all $(m,n) \neq (0,0)$.

Now set $P(w_1/2) = e_1$, $P(w_2/2) = e_2$ and $P(-(w_1 + w_2)/2) = e_3$. We have that e_i are roots of the cubic equation $x^3 - Ix - J = 0$.

Case 1: w_1, w_2 complex with $w_2 = \bar{w}_1$. In this case $e_2 = \bar{e}_1$ are complex roots so the discriminant Δ is negative.

Case 2: w_1 is real and w_2 is imaginary. In this case e_1 and e_2 are both real so the discriminant will be positive.

Figure 10

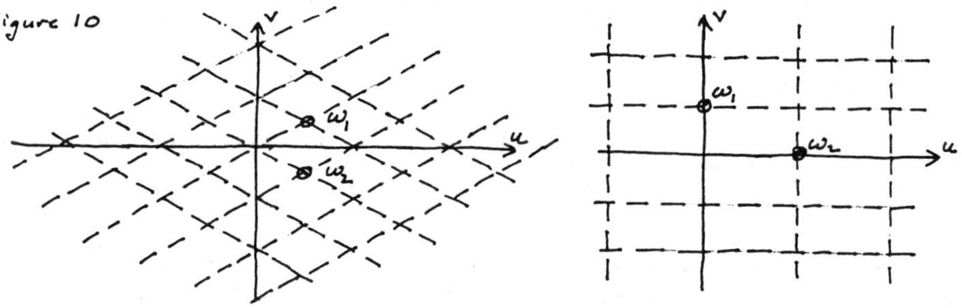

There is also an explicit expression for g(w) in the general case. We know $F(w) = b - 4P(w)$ where $b^3 - 4Ib - 16J = 1$ (b real). Let $b = 4P(a)$. By substitution and using the identity $(P')^2 = 4P^3 - IP - J$ we find that $16P'(a)^2 = 1$.

Note: Consider P(t), t real, $0 < t < \hat{w}$ where $\hat{w}$ is the first positive pole. We have $P'(\hat{w}/2) = 0$ and $P'(t) < 0$ for $0 < t < \hat{w}/2$. There is a value a, $0 < a < \hat{w}/2$ with $P'(a) = -1/4$.

$$\int_0^w [1/F(u)]du = \int_0^w \frac{du}{4P(a) - 4P(u)} = -\frac{1}{4}\int_0^w \frac{du}{P(u) - P(a)}$$

We now use formula #1036.02 in [3]. After taking exponentials we find

$$g(w) = Me^{2\zeta(a)w}\left[\frac{\sigma(a - w)}{\sigma(a + w)}\right]. \tag{4.27}$$

where $\zeta(w)$, $\sigma(w)$ are the standard Weierstrass Functions.

V. MINIMAL SURFACES IN H^3

We shall use the following model for hyperbolic 3-space.

(5.1) $$H^3 = \{x = (x_o, x_1, x_2, x_3) \in R^4 \,|\, \langle x,x\rangle = -1,\ x_o > 0\}$$

where we have set

(5.2) $$\langle x,y\rangle = -x_o y_o + \sum_{i=1}^{3} x_i y_i .$$

Let x(u,v) be a minimal surface conformally immersed in H^3 parameterized by its lines of curvature with fundamental forms as developed in Section II.

(5.3) $$ds^2 = e^{2\omega}(du^2 + dv^2) = E(du^2 + dv^2)$$

$$II = -du^2 + dv^2$$

with principal curvatures $k_1 = -e^{-2\omega}$, $k_2 = e^{-2\omega}$ where $\omega(u,v)$ satisfies the Gauss Equation (2.15)

(5.4) $$\Delta\omega = e^{2\omega} + e^{-2\omega} .$$

The differential equations determining the immersion into $H^3 \subset R^4$ are

(5.5) $$\begin{aligned} x_{uu} &= \omega_u x_u - \omega_v x_v + L\xi + e^{2\omega} x \\ x_{uv} &= \omega_v x_u + \omega_u x_v + M\xi \\ x_{vv} &= -\omega_u x_u + \omega_v x_v + N\xi + e^{2\omega} x \\ \xi_u &= (-L/E)x_u + (-M/E)x_v,\quad \xi_v = (-M/E)x_u + (-N/E)x_v . \end{aligned}$$

The tangent space at $x_o \varepsilon H^3$ consists of those vectors $y \varepsilon R^4$ with $\langle x_o, y \rangle = 0$. The unit normal ξ is determined by the condition that $\{e_1, e_2, \xi\}$ be an oriented orthonormal basis to the tangent space at x(u,v), where $e_1 = x_u/|x_u|$ and $e_2 = x_v/|x_v|$.

The fact that the system has a solution for x(u,v) unique up to an isometry of H^3 is an extension of O. Bonnet's theorem [11].

We recall the following facts regarding the geometry of planes and spheres in our model of H^3.

I.) A "plane" in H^3 is the intersection of $H^3 \subset R^4$ with a hyperplane of R^4 passing through the origin.

(5.6) $$\Pi = \{x \ \varepsilon \ H^3 \mid \langle m,x \rangle = 0\}$$

where $m = (m_o, m_1, m_2, m_3)$ satisfies $\langle \mathbf{m},\mathbf{m} \rangle > 0$ if Π is not empty.

II.) A "sphere" S[m,q] in H^3 where $\mathbf{m} \varepsilon R^4$ and q is a real number, is the set of points

(5.7) $$S[\mathbf{m},q] = \{x \ \varepsilon \ H^3 \text{ with } \langle \mathbf{m}, x \rangle = q\}$$

assuming the set is non-empty. As sphere is one of three types.

A) Closed (compact) Spheres:

These genuine spheres arise when $\langle \mathbf{m}, \mathbf{m} \rangle$ is negative. In this case one may choose $\langle \mathbf{m},\mathbf{m} \rangle = -1$. In order for the sphere to be non-empty it is necessary that $q < -1$. S[m,q] is then a compact sphere with center **m** (if $\langle \mathbf{m},\mathbf{m} \rangle = -1$). The radius R of this sphere satisfies the relation coshR = -q.

B) Horospheres:

A horosphere is determined by the set S[m,q] when $\langle \mathbf{m},\mathbf{m} \rangle = 0$, $\mathbf{m} \neq 0$ and $q < 0$.

C) Open Pspeudo-spheres:
Open pseudo-spheres are sets of the form S[**m**,q] where ⟨**m**,**m**⟩ is positive and q is arbitrary. Both horospheres and pseudospheres are unbounded sets in H^3. Also note that a plane is a pseudosphere obtained when q = 0.

<u>Theorem 5.1</u>: let x(u,v) be a minimal immersion of Joachimsthal type in H^3 such that for each $\hat{v}$, the line of curvature $x(u,\hat{v})$ is planar. Then $\omega(u,v)$ satisfies the additional equation

$$\omega_{uv} + \omega_u\omega_v = 0, \quad (e^{\omega}\omega_u)_v = (e^{\omega}\omega_v)_u = 0 \tag{5.8}$$

which implies that there exist functions a(u), b(v) for which $\omega_u = a(u)e^{-\omega}$ and $\omega_v = b(v)e^{-\omega}$.

A proof can be modelled on the Euclidean proof as presented by R. Walter [13]. It also follows from the following result.

<u>Theorem 5.2</u>: Let x(u,v) be a minimal immersion of Enneper type in H^3 such that for each $\hat{u}$, the curvature line $x(\hat{u},v)$ lies on a sphere $S[\mathbf{m}(\hat{u}),\, q(\hat{u})]$. (We allow all types of spheres.) There exist functions $\alpha(u)$, $\beta(u)$ such that

$$2\omega_u = \alpha(u)e^{\omega} + \beta(u)e^{-\omega}\,. \tag{5.9}$$

If $\theta(\hat{u})$ is the angle of intersection of the surface with the sphere along the curve $x(\hat{u},v)$ then

$$\alpha(u) = 2q/|N|\sin\theta \tag{5.10}$$

$$\beta(u) = 2\cot\theta$$

where $N = \mathbf{m} + \langle\mathbf{m},x\rangle x$, $|N|^2 = \langle\mathbf{m},\mathbf{m}\rangle + q^2$.

Proof: We are given $\mathbf{m}(u)$, $q(u)$ so that $x(u,v) \subset S[\mathbf{m}(u),q(u)]$ where $\mathbf{m} \neq 0$. We know that $\langle x,x\rangle = -1$ and $\langle \mathbf{m},x\rangle = q$. Differentiate with respect to v and we find $\langle \mathbf{m},x_v\rangle = 0$ and $\langle \mathbf{m},x_{vv}\rangle = 0$. Now write

$$\mathbf{m} = Ax_u + Bx_v + C\xi + Dx$$

where $\{x_u, x_v, \xi, x\}$ is the Darboux frame. We see at once that $B = 0$ and $D = -q(u)$.

If x is a point on the sphere $S[\mathbf{m},q]$ then the tangent space to $S[\mathbf{m},q]$ at x consists of those vectors $T \in R^4$ for which $\langle x,T\rangle = 0$ and $\langle \mathbf{m},T\rangle = 0$. We claim that $N = \mathbf{m} + \langle \mathbf{m},x\rangle x$ is a normal vector to the sphere $S[\mathbf{m},q]$ at x. First $\langle N,x\rangle = 0$ so that $N \in T_x(H^3)$. Secondly, $\langle N,T\rangle = 0$ for all tangent vectors to the sphere. Moreover $\langle N,N\rangle = \langle \mathbf{m},\mathbf{m}\rangle + q^2$ is positive.

Since ξ is the unit normal to the surface at x we have $\langle \xi,N\rangle = |N|\cos\theta = \langle \mathbf{m},\xi\rangle = C(u)$. Finally to calculate A, we use

$$\langle N,e_1\rangle = |N|\sin\theta = e^{-\omega}\langle \mathbf{m},x_u\rangle = e^{\omega}A.$$

Now use (5.5), the fact that $\langle \mathbf{m},x_{vv}\rangle = 0$ and our known values for A, B, C and one finds

$$\omega_u = \left(\frac{q}{|N|\sin\theta}\right)e^{\omega} + (\cot\theta)e^{-\omega}. \tag{5.11}$$

Q.E.D.

From (5.11) we observe that if $q(\hat{u}) = 0$ then the curvature line $x(\hat{u},v)$ is planar and if $q(u) \equiv 0$ the surface is of Joachimsthal type with (5.8) being valid.

Theorem 5.3: Let $x(u,v)$ be a minimal immersion of Enneper type in H^3 where $\omega(u,v)$ satisfies the Gauss equation (5.9) and the supplementary equation (5.11). The sphere $S[\mathbf{m}(u),q(u)]$ is a compact sphere if $\alpha^2 - \beta^2 > 4$, it is a horosphere if $\alpha^2 - \beta^2 = 4$, and it is a pseudosphere if $\alpha^2 - \beta^2 < 4$. Also if

$\alpha^2 - \beta^2 > 4$ the center of the sphere is given by

$$(5.12)\quad \mathbf{m}(u) = \frac{\varepsilon}{\sqrt{\alpha^2 - \beta^2 - 4}}\left[2e^{-\omega} x_u + \beta\xi - \alpha x\right],\ \varepsilon = -\mathrm{sign}(\alpha).$$

<u>Proof</u>: From (5.11) we have $\alpha(u) = 2q/|N|\sin\theta$, $\beta(u) = 2\cot\theta$ where $|N|^2 = \langle\mathbf{m},\mathbf{m}\rangle + q^2$. Suppose first that $\langle\mathbf{m},\mathbf{m}\rangle < 0$ so that $S[\mathbf{m},q]$ is a compact sphere. We may assume $\langle\mathbf{m},\mathbf{m}\rangle = -1$ and $q < -1$ so $|N|^2 = q^2 - 1$. In this case we find

$$(5.13)\qquad \alpha^2 - \beta^2 = 4\left(1 + \frac{1}{(q^2 - 1)\sin^2\theta}\right) > 4.$$

If $S[\mathbf{m},q]$ is a horosphere with $\langle\mathbf{m},\mathbf{m}\rangle = 0$, $\mathbf{m} \neq 0$ and $q < 0$, then $|N|^2 = q^2$ and $\alpha^2 - \beta^2 = 4$. Finally if $S[\mathbf{m},q]$ is a pseudo-sphere with $\langle\mathbf{m},\mathbf{m}\rangle > 0$ we may assume $\langle\mathbf{m},\mathbf{m}\rangle = 1$ and $|N|^2 = q^2 + 1$. We now find

$$(5.14)\qquad \alpha^2 - \beta^2 = 4\left(1 - \frac{1}{(q^2 + 1)\sin^2\theta}\right) < 4.$$

From the proof of Theorem 5.2 we have $\mathbf{m} = Ax_u + Bx_v + C\xi + Dx$ where $e^{\omega}A = |N|\sin\theta$, $B = 0$, $C = |N|\cos\theta$, $D = -q$ where $\langle\mathbf{m},\mathbf{m}\rangle = -1$ and $\langle x\cdot\mathbf{m}\rangle = q$, $q < -1$. Also from (5.11) we see that $\alpha\sin\theta$ is negative. We use (5.13) and the fact that $|N|^2 = q^2 - 1$ to conclude $Ae^{\omega} = |N|\sin\theta = \varepsilon\left[2/\sqrt{\alpha^2 - \beta^2 - 4}\right]$ where $\varepsilon = -\mathrm{sign}(\alpha)$ $C = |N|\cos\theta = (\varepsilon B)/\sqrt{\alpha^2 - \beta^2 - 4}$ and $D = -q = -\alpha|N|\sin\theta/2$ $= -\varepsilon\alpha/\sqrt{\alpha^2 - \beta^2 - 4}$. This gives us the desired formula for m.

Q.E.D.

We now apply Theorem 2.3. If $\omega(u,v)$ is a solution to the Gauss equation (5.4) where $\omega_v \not\equiv 0$ and if $2\omega_u = \alpha(u)e^{\omega} + \beta(u)e^{-\omega}$

then $\alpha(u)$, $\beta(u)$ are solutions to the system (2.18) where $A = -1$, $B = 1$.

$$\begin{aligned} \alpha'' &= a\alpha - 2\alpha^2\beta + 2\beta \\ \beta'' &= a\beta - 2\alpha\beta^2 - 2\alpha \end{aligned} \tag{5.15}$$

as well as the condition (2.19).

Theorem 5.4: Let $x(u,v)$ be a conformal minimal immersion into H^3 of Enneper type where $\omega_v \not\equiv 0$. Suppose that on some interval the spheres $S[\mathbf{m}(u),q(u)]$ are compact with $\langle \mathbf{m},\mathbf{m}\rangle = -1$, $q < -1$, with centers $\mathbf{m}(u)$ given by (5.12). The centers all lie on a geodesic of H^3.

Proof: Let $\mathbf{f}(u) = 2e^{-\omega}\mathbf{x}_u + \beta\xi - \alpha\mathbf{x} \in R^4$. We claim that $\mathbf{f}''(u) = \varphi(u)\mathbf{f}(u)$ for some scaler function $\varphi(u)$. Supposing this to be the case, choose two linearly independent vectors $\mathbf{a}$, $\mathbf{b} \in R^4$ with $\langle \mathbf{a},\mathbf{a}\rangle = \langle \mathbf{b},\mathbf{b}\rangle = 1$ satisfying the conditions $\langle \mathbf{a},\mathbf{f}(u_o)\rangle = \langle \mathbf{b},\mathbf{f}(u_o)\rangle = \langle \mathbf{a},\mathbf{f}'(u_o)\rangle = \langle \mathbf{b},\mathbf{f}'(u_o)\rangle = 0$ for some u_o. Let $F(u) = \langle \mathbf{a},\mathbf{f}(u)\rangle$ and $G(u) = \langle \mathbf{b},\mathbf{f}(u)\rangle$. We have $F(u_o) = F'(u_o) = 0$ and $F''(u) = \varphi(u)F(u)$. By the existence and uniqueness theorem for O.D.E.'s we conclude that $F(u) \equiv 0$ and similarly $G(u) \equiv 0$ or $\langle \mathbf{a},\mathbf{f}(u)\rangle = \langle \mathbf{b},\mathbf{f}(u)\rangle = 0$. Since $\mathbf{m}(u)$ is just a scaler multiple of $\mathbf{f}(u)$ we see that $\mathbf{m}(u) \in H^3$ and lies on the line ℓ given by the equation

$$\ell = \{x \in H^3 \mid \langle \mathbf{a},\mathbf{x}\rangle = \langle \mathbf{b},\mathbf{x}\rangle = 0\}.$$

The proof that $\mathbf{f}''(u) = \varphi(u)\mathbf{f}(u)$ is a tedious but straightforward calculation (which we omit). One uses equations (5.5) and (5.15).
Q.E.D.

We now wish to solve the Gauss equation (5.4) subject to the auxillary condition (5.9)where $\{\alpha(u), \beta(u)\}$ are solutions to the system (5.15). As earlier, if we set $X = e^{\omega}$ then (2.19) and (2.22) are

$$(5.16)\quad 2X_u = \alpha(u)X^2 + \beta(u)$$

$$4X_v^2 = (4 - \alpha^2)X^4 - 4\alpha' X^3 + 6\gamma X^2 + 4\beta' X - (4 + \beta^2)$$

$$\text{where } 6\gamma = 6\alpha\beta - 4a$$

We note the surprising fact.

<u>Theorem</u> <u>5.5</u>: The only minimal immersions of Joachimsthal type in H^3 are the surfaces of revolution.

<u>Proof</u>: If $\omega_v \not\equiv 0$ and $x(u,v)$ is of Joachimsthal type then $\alpha(u) \equiv 0$ in (5.9). However if $\alpha(u) \equiv 0$ then $\beta(u) \equiv 0$ as well using (5.15). Therefore $\omega(u,v)$ is a function of v alone.

Q.E.D.

We now use the separation of variables method as discussed in Section II. The change of variables formula (2.24) is awkward and we prefer the following formula which differs from (2.24) by replacing s by s - 2i and t by t - 2i. We set

$$(5.17)\qquad \alpha\beta = s + t$$

$$\alpha^2 - \beta^2 = 4 - st.$$

One has immediately the identities,

$$(5.18)\qquad (\alpha - i\beta)^2 = -(s + 2i)(t + 2i)$$

$$(\alpha + i\beta)^2 = -(s - 2i)(t - 2i)$$

as well as the explicit formulae

$$(5.19)\qquad 2\alpha^2 = (4 - st) + \sqrt{4(s - t)^2 + (st + 4)^2}$$

$$2\beta^2 = -(4 - st) + \sqrt{4(s - t)^2 + (st + 4)^2}$$

$$2s = \alpha\beta + \sqrt{(\alpha^2 - 4)(\beta^2 + 4)}$$

$$2t = \alpha\beta - \sqrt{(\alpha^2 - 4)(\beta^2 + 4)}.$$

If $\alpha^2 > 4$ then s and t are real $s \neq t$ and (5.17) determines a diffeomorphism of the region $\alpha > 2$ ($\alpha < -2$) in the $\alpha - \beta$ plane onto the region $s > t$ in the $s - t$ plane (we select $s > t$). If $\alpha^2 < 4$ then s and t are complex with $\bar{s} = t$, $Im(s) \neq 0$. We choose s by requiring that $Im(s) > 0$. In this case the equations (5.17) determine a double cover of the strip $\alpha^2 < 4$ onto the upper half s-plane, since (α, β) and $(-\alpha, -\beta)$ are mapped into the same point. Note that $(\alpha, \beta) = (0, 0)$ is mapped into $(s, t) = (2i, -2i)$. Finally, if $\alpha^2 = 4$ then $s = t$.

Besides satisfying the properties listed in Section II the solutions $\omega(u,v)$ to (5.4)(5.9) which we construct here will have the following property. There will be a maximal domain Ω for $\omega(u,v)$ with limit $\omega(u,v) = +\infty$ at the boundary of Ω. This domain will not be simply connected. It will be contained in a vertical strip (usually of finite width) and the domain will have holes which are repeated periodically in the v-direction. The corresponding immersion $x(u,v)$ can be integrated as a single-valued mapping from Ω into H^3.

The system (5.15) has the integral

$$\alpha'\beta' = a\alpha\beta - \alpha^2\beta^2 + \beta^2 - \alpha^2 + h = \varphi(\alpha,\beta) + h$$

with corresponding Hamiltonian

$$\mathcal{H}(\alpha,\beta,p_\alpha,p_\alpha) = p_\alpha p_\beta - \varphi(\alpha,\beta) = h.$$

The transformation (5.17) separates the variables in the Hamilton-Jacobi equation leading to the system

$$(5.20) \qquad s'(\lambda)^2 = (s^2 + 4)\, g(s)$$

$$t'(\lambda)^2 = (t^2 + 4)g(t), \qquad \lambda'(u) = 2/(s - t)$$

where now

$$(5.21) \qquad g(s) = -s^3 + as^2 + (h - 4)s + k.$$

One checks that

$$4g(s) = 4(-s + \frac{a}{3})^3 - I(-s + \frac{a}{3}) + J$$

$$I = -16 + 4h + (4a^2/3)$$

$$J = 4k + (4a/3)(h - 4) + (8a^3)/27$$

$$4k = (\alpha\beta' - \alpha'\beta)^2 + 4(\alpha'^2 - \beta'^2) + 4(\alpha\beta - a)(\alpha^2 - \beta^2 - 4).$$

Here I and J are invariants for the polynomials $p(u,X)$ in (5.16).

The solution of (5.15) splits into two cases. If $|\alpha| > 2$ then s and t are real with $s > t$ while if $|\alpha| < 2$ then s and t are complex with $\bar{s} = t$ and $\mathrm{Im}(s) > 0$. If $|\alpha(\hat{u})| > 2$ then by considering (5.16) we see that $\omega(\hat{u},v)$ is bounded and periodic for all v while if $|\alpha(\hat{u})| \leq 2$ the solution $\omega(\hat{u},v)$ will become positively infinite and the corresponding immersion $x(\hat{u},v)$ will become unbounded.

Case A: $\Delta = I^3 - 27J^2 < 0$. Here $g(s)$ has exactly one real root, $g(s) = -(s - s_o)h(s)$ where $h(s)$ is positive for all s. In this case the polynomial $p(u,X)$ in (5.16) will have eactly two roots if $\alpha^2 \neq 4$ and one root if $\alpha^2 = 4$.

I) $|\alpha| > 2$ so that $s > t$.

The solutions $s(\lambda)$, $t(\lambda)$ solving (5.20) will be congruent with $t(\lambda) = s(\lambda - \lambda_o)$ for some λ_o. If we choose initial conditions so that $s(0) = 0$ then $s(\lambda)$ will have a half-period M with limit $s(\lambda) = -\infty$ as λ approaches M.

$$M = \int_0^\infty \frac{ds}{\sqrt{(s^2+4)g(s)}} < +\infty$$

Figure 11

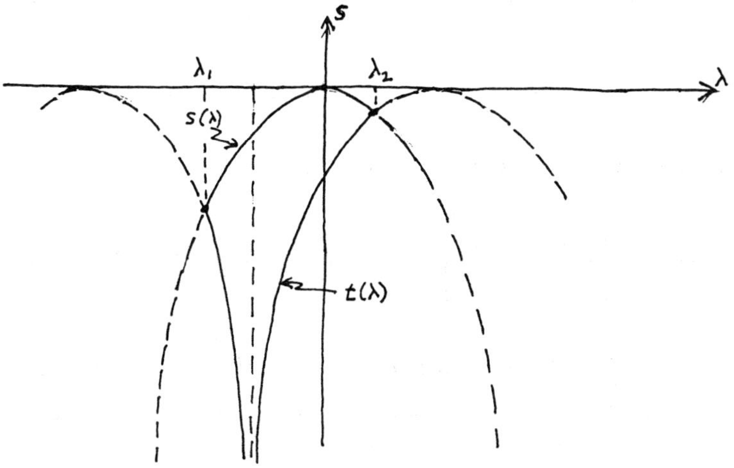

We have s > t on an interval $\lambda_1 < \lambda < \lambda_2$ with $s(\lambda_1) = t(\lambda_1) = s_1$ and $s(\lambda_2) = t(\lambda_2) = s_2$. There is a value λ_3 in the interval with $t(\lambda_3) = -\infty$. The relation $\lambda'(u) = 2/(s-t)$ transforms the finite interval (λ_1,λ_2) into a finite interval (u_1,u_2) with $s(u_1) = s_1$, $s(u_2) = s_2$, $s(u_3) = -\infty$.

a) Near $u = u_3$ we have $\alpha'(u) < 0$ with $\alpha(u_3-) = -\infty$ and $\alpha(u_3+) = +\infty$. Also $\alpha\beta = s + t$ is negative so that $\beta'(u) > 0$ for u near u_3, $\beta(u_3-) = +\infty$, $\beta(u_3+) = -\infty$.

Furthermore, we find that for $u = u_3$

$$(5.22) \qquad X^2(u_3,v) = e^{2\omega(u_3,v)} = \frac{-s_3 + \sqrt{s_3^2 + 4}}{2}$$

where $s_3 = s(u_3)$.

Proof of (a): From the formulae (5.17) one finds the following limits as $t \to -\infty$.

$$\text{limit}\ \frac{\alpha^2}{(-t)} = \frac{s + \sqrt{s^2 + 4}}{2} \qquad \text{limit}\ \frac{\beta^2}{(-t)} = \frac{-s + \sqrt{s^2 + 4}}{2}$$

$$\text{limit}\ \frac{\alpha^2 + \beta^2}{(-t)} = \sqrt{s^2 + 4} \qquad \text{limit}\ \frac{\alpha\beta}{(-t)} = -1$$

By considering the derivatives, one finds

$$\text{limit}\ \left|\frac{\alpha'(u)}{(-t)}\right| = \left(\frac{s + \sqrt{s^2+4}}{2}\right)^{1/2}, \quad \text{limit}\ \left|\frac{\beta'(u)}{(-t)}\right| = \left(\frac{-s + \sqrt{s^2+4}}{2}\right)^{1/2}.$$

Set $m^2 = (-s + \sqrt{s^2 + 4})/2$ and $n^2 = (s + \sqrt{s^2 + 4})/2$ where m,n are positive $mn = 1$. Since α and β are of opposite signs we have $\alpha' > 0$, $\beta' < 0$ or the reverse. If $\alpha' < 0$ and $\beta' > 0$ we find using (5.16).

$$\text{limit}\ \frac{p(u,X)}{(-t)} = -\frac{1}{m^2}(X - m)^4$$

while for $\alpha' > 0$, $\beta' < 0$ the same limit is $-(X + m)/m^2$. For u near u_3 the roots of $p(u,X)$ are positive and so $\alpha' < 0$ and (5.22) follows.

b) Suppose for $u = \hat{u}$ we have $s(\hat{u}) = t(\hat{u}) = c$ so that $|\alpha(\hat{u})| = 2$. We find

(5.23) a) $|\alpha'(\hat{u})| = 2\sqrt{g(c)} \,/\, \sqrt{4 + c^2}$

b) $\beta'(\hat{u}) = \varepsilon \left[\sqrt{c^2 + 4}\ g'(c)/2\sqrt{g(c)}\right]$, $\varepsilon = \operatorname{sgn}\alpha'(\hat{u})$.

These formulae follow by differentiating the change of variables formula (5.17) and using (5.20).

We have $s > t$ for $u_1 < u < u_2$ determining the solutions $\langle\alpha(u), \beta(u)\rangle$ or $\langle-\alpha(u), -\beta(u)\rangle$ to (5.15). The correct choice is governed by the condition that the two real roots of $p(u, X)$ be positive. It follows that the cubic polynomials $p(u_1, X)$ and $p(u_2, X)$ must have one positive root. This will be the case if $\alpha'(u_i)$ is negative. Therefore $\alpha(u_1) = -2$ and $\alpha(u_2) = +2$. The trajectory $\langle\alpha(u), \beta(u)\rangle$, $u_1 < u < u_2$ follows the diagrammed sketch with $\alpha(u_3)$ infinite.

Figure 12

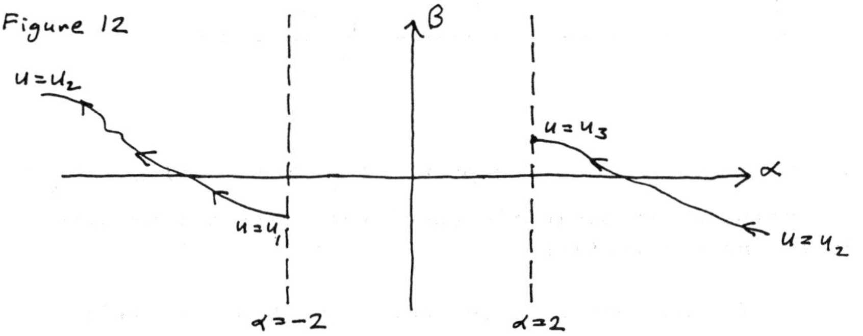

From the construction we observe that the initial and final values s_1, s_2 at u_1, u_2 are inversely related and that there is exactly one value s^* with $s_1 = s_2 = s^*$.

The behavior of $\omega(u,v)$ $u_1 \le u \le u_2$ is indicated in the following sketch.

Figure 13

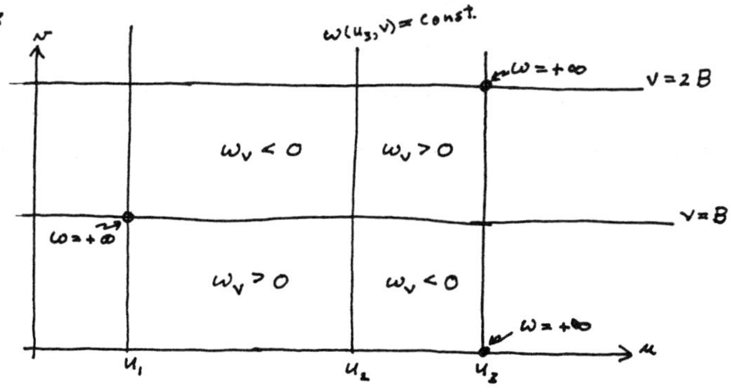

The solution $\omega(u,v) = +\infty$ at the points (u_1, kB) k odd and (u_2, kB) k even. The solution is symmetric about the lines $v = kB$ with period $2B$ in the v-direction.

II. $|\alpha| < 2$ so that s,t are complex with $\bar{s} = t$, $\text{Im}(s)$ positive.

Since $\lambda'(u) = 2/(s - t) = -i/\text{Im}(s)$ it follows that $\lambda(u) = \lambda_0 - i\,\varepsilon(u)$ where $\varepsilon'(u) > 0$. We replace the parameter λ by ε and (5.20) becomes

$$(5.24) \qquad \begin{aligned} s'(\varepsilon)^2 &= -(s^2 + 4)g(s) \\ t'(\varepsilon)^2 &= -(t^2 + 4)g(t) \end{aligned} \qquad \varepsilon'(u) = \frac{-2i}{s - \bar{s}} > 0$$

where $\text{Im}(s) > 0$. The differential equation for $s(\varepsilon)$ determines a line field on $\text{Im}(s) > 0$ given by

(5.25) $$\operatorname{Im}\left[\frac{ds^2}{(s^2+4)g(s)}\right] = 0$$

$$\operatorname{Re}\left[\frac{ds^2}{(s^2+4)g(s)}\right] < 0$$

with singularities at $s = 2i$ and at any zeros of $g(s)$ in the upper half plane. From the equations (5.17)

$$\alpha\beta = s + \bar{s}$$

$$\alpha^2 - \beta^2 = 4 - |s|^2 \qquad \text{when } \bar{s} = t$$

we see that $\alpha = 0$ when $s = ib$ where $b > 2$ while $\beta = 0$ when $s = ib$, $0 < b < 2$. Below is a sketch of a typical line field when $g(s)$ has one real root. I picked $g(s) = -s(s^2 + 1)$.

Figure 14

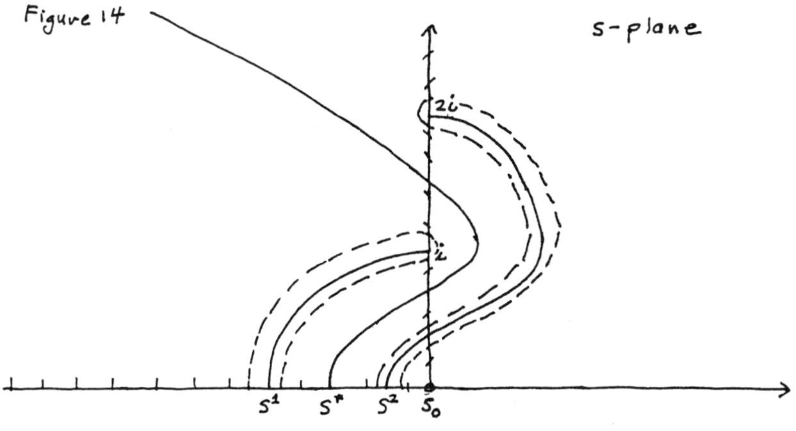

The lines are integral curves to (5.24) and may be traversed in either direction. For the example sketched we find

a) For $s > s_o$ (where $g(s_o) = 0$) the line field is horizontal while for $s < s_o$ the field is vertical. All integral curves enter the region $Im(s) > 0$ along the half line $s < s_o$.

b) All integral curves start and finish on the line $Re(s) < s_o$, $Im(s) = 0$ except for the following curves.

i) One curve starts at a value $s = s^1$ and terminates at $s = i$ (or more generally at the zero of $g(s)$).

ii) One curve starts at a value $s = s^2$ and terminates at $s = 2i$.

iii) One curve starts at $s = s*$ and becomes asymptotic to the ray $\arg(s) = 2\pi/3$.

All curves, including the one of infinite length are traversed in finite ε-time and also finite u-time. The proof of there assertions follow from the equations (5.24).

We now consider what happens to the corresponding curves in the (α,β)-plane.

c) Suppose $\mathscr{C}$ is an integral curve of (5.25) which starts at a value $s_1 < s^1$ and terminates at a value s_2 with $s^1 < s_2 < s*$. This curve will not cross the imaginary axis at a point above $s = 2i$. Consequently α is never zero so that $0 < |\alpha| \le 2$. On the other hand $\beta = 0$ twice and the corresponding curve in the $\alpha - \beta$ plane will start and end either on the line $\alpha = 2$ or $\alpha = -2$.

Figure 15

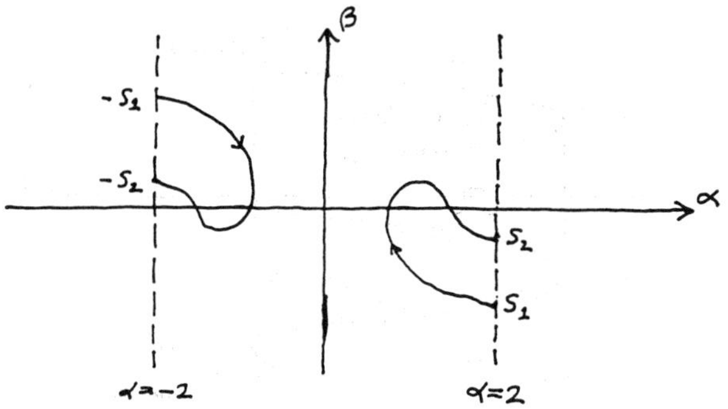

Note: $\alpha\beta = 2s$ (if s is real). Therefore either $\alpha = 2$, $\beta = s$ or $\alpha = -2$, $\beta = -s$.

Consider the case where $\alpha(u_1) = \alpha(u_2) = 2$ where $u_1 < u_2$ and $\beta(u_1) = s_1$, $\beta(u_2) = s_2$. It follows that $\alpha'(u_1) < 0$ hence the cubic polynomial $p(u_1,X)$ (5.16) has a single positive root so that $X(u_1,v)$ (resp. $\omega(u_1,v)$) becomes positively infinite when $v = kB$ (k odd). For $u_1 < u < u_2$ the quartic polynomial $p(u,X)$ will still have a single positive root and the coefficient of X^4 is positive. This will cause the graph of $X(u,v)$ as a function of v to blow up in a time less than B thus creating a hole in the domain for $\omega(u,v)$.

Finally as $u \to u_2$ we see that $p(u_2,X)$ is again a cubic polynomial where $\alpha'(u_2)$ is now positive. This means that $p(u_2,X)$ is negative for all positive X. There is no solution for $\omega(u_2,v)$ and the line $u = u_2$ is a barrier for the domain of ω.

Figure 16

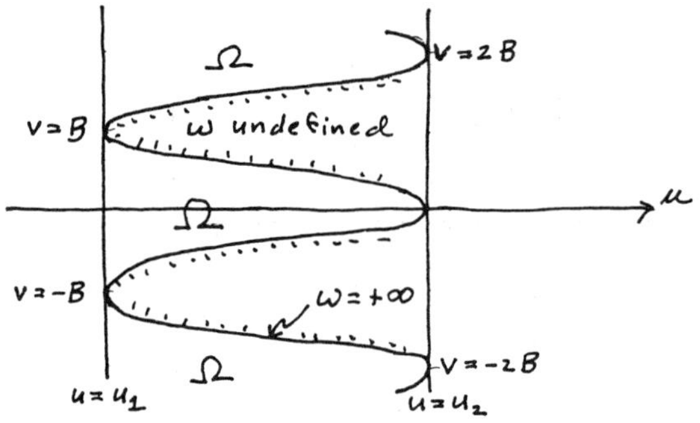

Similarly if we consider the curve $(\alpha(u), \beta(u))$ $u_1 \leq u < u_2$ which starts and ends on the line $\alpha = -2$ then a left side barrier $u = u_1$ is created for the domain of $\omega(u,v)$.

If $\mathscr{C}$ is that curve which starts at s^1 goes to $s = i$ and returns to s^1 and returns to s^1 then a similar behavior is observed.

d) $\mathscr{C}^*$ is that curve starting at s^* becoming asymptotic to $\arg(s) = 2\pi/3$. In this case one finds that α^2 approaches one as s becomes infinite. also $\alpha\beta = 2\mathrm{Re}(s) < 0$ along $\mathscr{C}^*$, thus either $\alpha \to 1$, $\beta - -\infty$ or $\alpha \to -1$, $\beta \to +\infty$. In our particular example we see that α is never 0 while β vanishes twice.

Figure 17

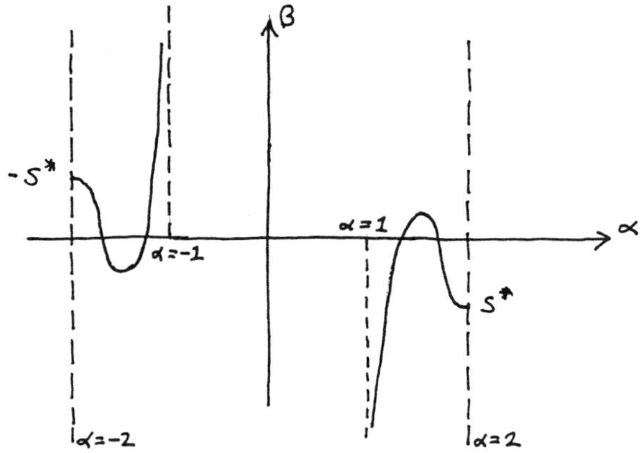

The curve is traversed in finite u-time $u_1 \leq u \leq u_2$. If $\alpha(u_1) = 2$ then $\alpha'(u_1) < 0$ and as in the previous case the cubic polynomial $p(u_1, X)$ has a single positive root. As u approaches u_2 we have $\alpha \to 1$ and $\beta \to -\infty$. It follows that on any interval $[0, X_0]$ the polynomials $p(u,X)$ will be negative for u near u_2. Once again we have a barrier at $u = u_2$ and the maximal domain for $\omega(u,v)$ lies to the left of this line.

e) Let $\mathcal{C}$ be an integral curve of (5.25)which starts at a value s_1, $s^* < s_1 < s^2$ and terminates at s_2, $s^2 < s_2 < s_0 = 0$. Here $\alpha(u)$ has exactly one zero on the interval $u_1 \leq u \leq u_2$. Since we want $\alpha'(u_1) < 0$ we choose the curve so that $\alpha(u_1) = 2$ and $\alpha(u_2) = -2$.

Figure 18

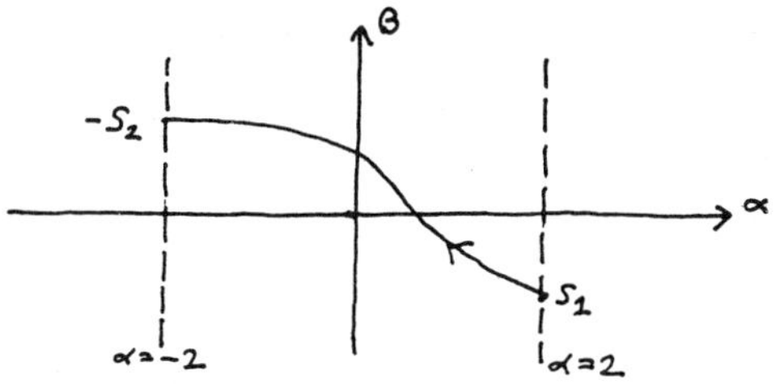

In this case there is no barrier. The domain for $\omega(u,v)$ has a hole starting at $u = u_1$, $v = B$ and closing up at $u = u_2$, $v = B$.

Figure 19

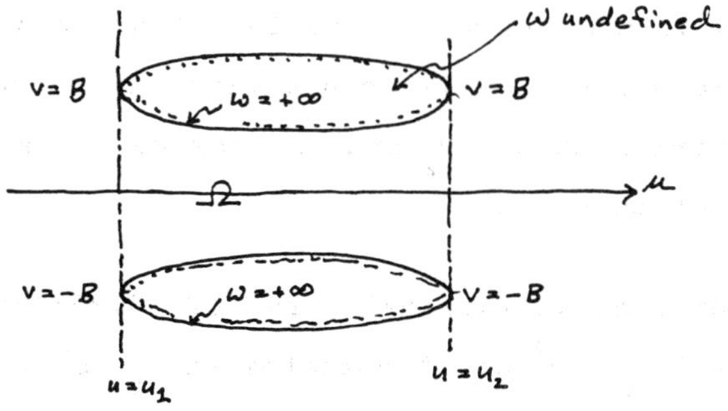

In general one obtains a maximal domain for $\omega(u,v)$ to be a region with periodic holes and which lies in a vertical strip $u_1 < u < u_2$ (u_1, u_2 perhaps infinite).

An especially interesting case arises when $g(s) = -(s - s_o)(s^2 + 4)$. Now the value $s = 2i$ is a repeated root for the system (5.24) and the line field (5.25) is somewhat different. The index of the line field at $s = 2i$ is one and the field appers as in the figure.

Figure 20

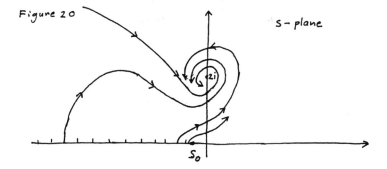

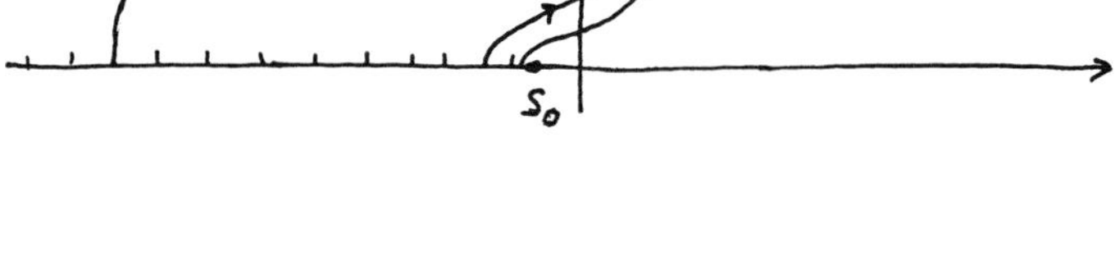

All trajectories spiral (in infinite u-time) towards $s = 2i$. One trajectory is asymptotic to $\arg(s) = 2\pi/3$ while the others terminate on the real axis, $\mathrm{Re}(s) < s_o$.

Now $s = 2i$ corresponds to the constant solution $\alpha(u) = \beta(u) = 0$. This means that $\omega_u \equiv 0$ so that ω is a function of v alone. If we solve (5.16) we find

$$4X_v^2 = 4(X^4 - 6s_oX^2 - 1), \quad X = e^{\omega}.$$

This gives us a one-parameter family of one-dimensional solutions. Referring to (5.11) one sees that for each $\hat{u}, x(\hat{u},v)$ lies on a plane and that the surface cuts this plane at right angles. These are generalized surfaces of revolution which are the analogues of Delaunay type surfaces in R^3. The domain for $\omega(u,v) = \omega(v)$ is a horizontal strip, $-M < v < M$ where M is found by the equation

$$M = \int_{x_o}^{\infty} \frac{dx}{\sqrt{x^4 - 6s_ox^2 - 1}}, \quad x_o^4 - 6s_ox_o^2 - 1 = 0, \quad x_o > 0$$

What about the solution $\omega(u,v)$ when $g(s) = -(s - s_o)(s^2 + 4)$? The domain for $\omega(u,v)$ will consist of three parts. There will be a middle part which is a vertical strip $u_1 \le u \le u_2$. Inside this strip $|\alpha| > 2$ and the solution $\omega(u,v)$ will be periodic in v(half period B) as analyzed earlier. For fixed $\hat{u}$, the curve $x(\hat{u},v)$ is a bounded curve in H^3.

At $u = u_1$ ($u = u_2$) we have $s = t = s_1$ or $s = t = s_2$ so that for $u < u_1$ ($u > u_2$) s and t are complex and $|\alpha| < 2$. Consider the case $u \ge u_2$. At $u = u_2$ we must have $\alpha = 2$, $\alpha'(u_2) < 0$ and the

$u \to +\infty$ the pair (α,β) spirals toward $(0,0)$ and the immersed pieces become of Delaunay type. The domain appears as

Figure 21

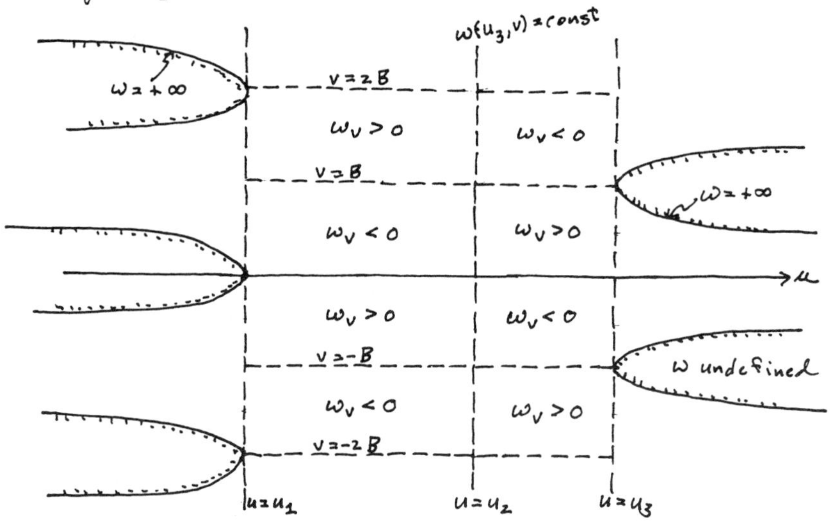

Case B: $\Delta = I^3 - 27J^2 > 0$. Here $g(s) = -(s-\hat{a})(s-\hat{b})(s-\hat{c})$ has three real roots $\hat{a} > \hat{b} > \hat{c}$. The analysis is similar to Case A but is in some regards more complicated. First consider the case $s > t$ with s,t real. Since $g(s)$ has three distinct real roots there are two types of solutions to (5.24) $s'(\lambda)^2 = (s^2 + 4)g(s)$. One solution is oscillatory and bounded with $\hat{a} \geq s(\lambda) \geq \hat{b}$ while the other solution satisfies $s(\lambda) \leq \hat{c}$ and possesses vertical aymptotes. From (5.20) we are to pick two solutions $s(\lambda)$, $t(\lambda)$ with $s > t$. If we choose $s(\lambda)$ to be the bounded solution and $t(\lambda)$ to be the unbounded solution than a straightforward calculation reveals that the polynomial $p(u,X)$ in (5.16) is negative for all real X. There are no solutions for $\omega(u,v)$.

Therefore $s(\lambda)$, $t(\lambda)$ must be congruent solutions $t(\lambda) = s(\lambda - \lambda_0)$ and the condition $s > t$ limits the interval of definition to be some finite interval $\lambda_1 \le \lambda \le \lambda_2$ $(u_1 \le u \le u_2)$.

Figure 22

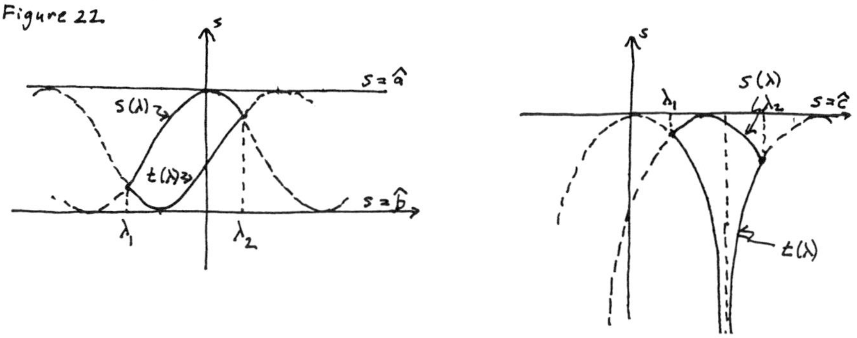

In case (i) the finite interval $\lambda_1 \le \lambda \le \lambda_2$ corresponds to a finite interval $u_1 \le u \le u_2$. On this interval either $\alpha(u) \ge 2$ or $\alpha(u) \le -2$. If $\alpha(u) > 2$ for $u_1 < u < u_2$ with $\alpha(u_1) = \alpha(u_2) = 2$ then $\alpha'(u_1) > 0$ and $\alpha'(u_2) < 0$.

For $u_1 < u < u_2$ the polynomial $p(u,X)$ will have 4 real roots, two of them positive.

Figure 23

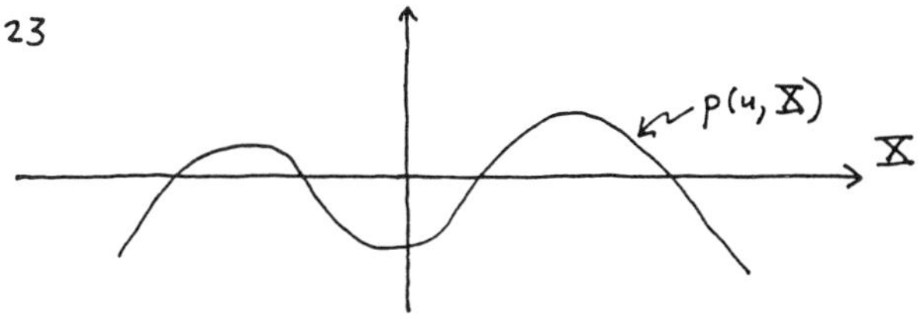

As the end points we will have

Figure 24

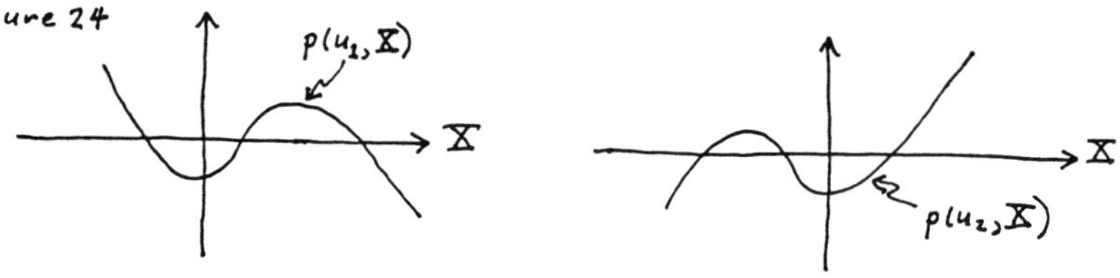

The solution $\omega(u,v)$ extends across the interval $u_1 \leq u \leq u_2$. At $u = u_2$ the solution $\omega(u_2,v)$ becomes infinite and develops a singularity at $v = B$, while at $u = u_1$, the solution $\omega(u_1,v)$ remains bounded.

For $s(\lambda)$, $t(\lambda)$ as in Figure (ii) the pair $(\alpha(u), \beta(u))$ behave as in the case where $g(s)$ has one real root. We need $\alpha(u_1) = 2$, $\alpha'(u_1) < 0$, $\alpha(u_2) = -2$, $\alpha'(u_2) < 0$ $\alpha(u_3) = \infty$. For $u \neq u_3$ the polynomials $p(u,X)$ will have 4 positive roots. At the end points $p(u_i,X)$ will have three positive roots.

Finally in the range $|\alpha| < 2$ we consider the line field determined by (5.25). In this case the line field appears as in the figure.

Figure 25

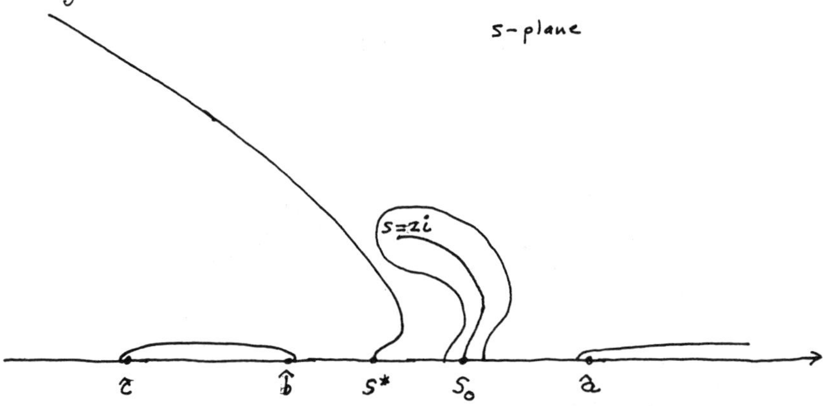

The integral curves can enter or leave the region on the intervals $-\infty < s < \hat{c}$ or $\hat{b} < s < \hat{a}$. There will be one curve connecting s_0 to $s = 2i$ while another curve connects $s = s^*$ to the asymptote $\arg(s) = 2\pi/3$. For suitable values these two special curves can connect up. We now can construct solutions $\omega(u,v)$ and the maximal domains as in Case A.

VI. ILLUSTRATIONS

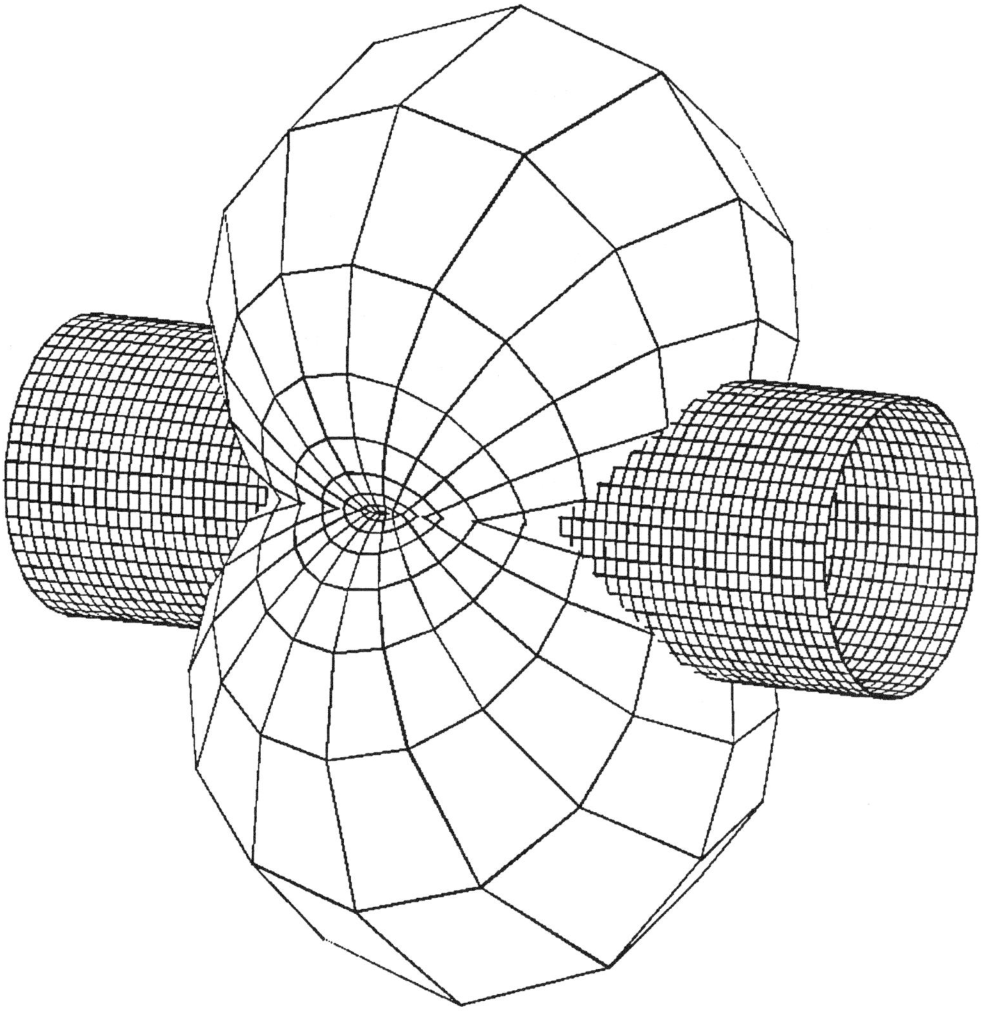

[839 by 806 pixels, 1.000 aspect ratio, 7.00 by 6.72 inches

1. A cmc immersed surface, H = 1/2, with embedded cylindrical ends and two lobes in the middle. It is of Joachimsthal type (see page 26).

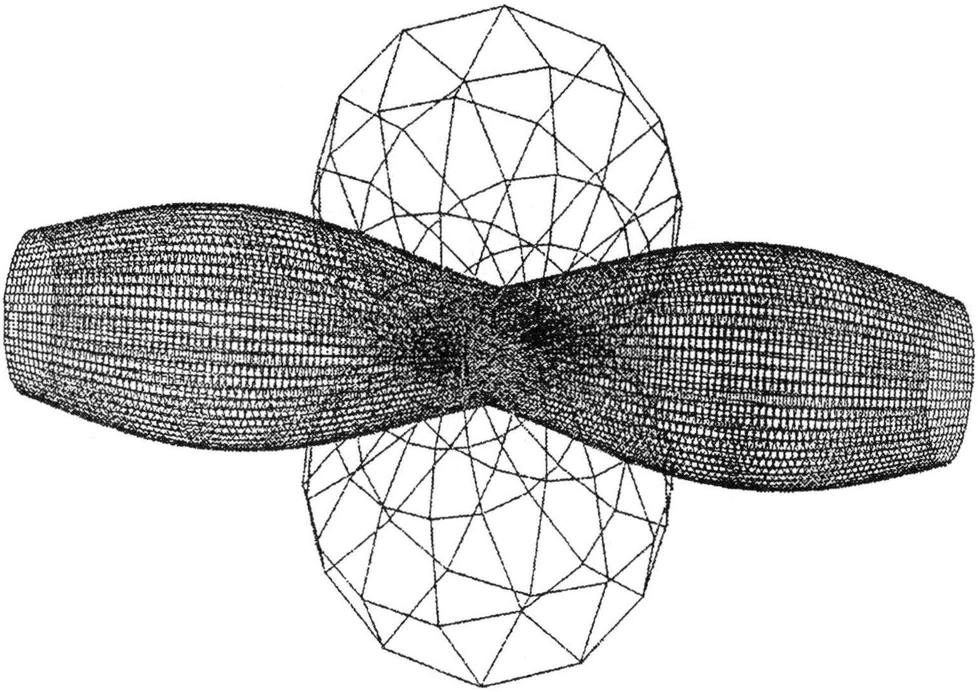

2. A cmc immersed surface, H = 1/2, with embedded unduloidal ends and two lobes. It is of Enneper type (see page 29).

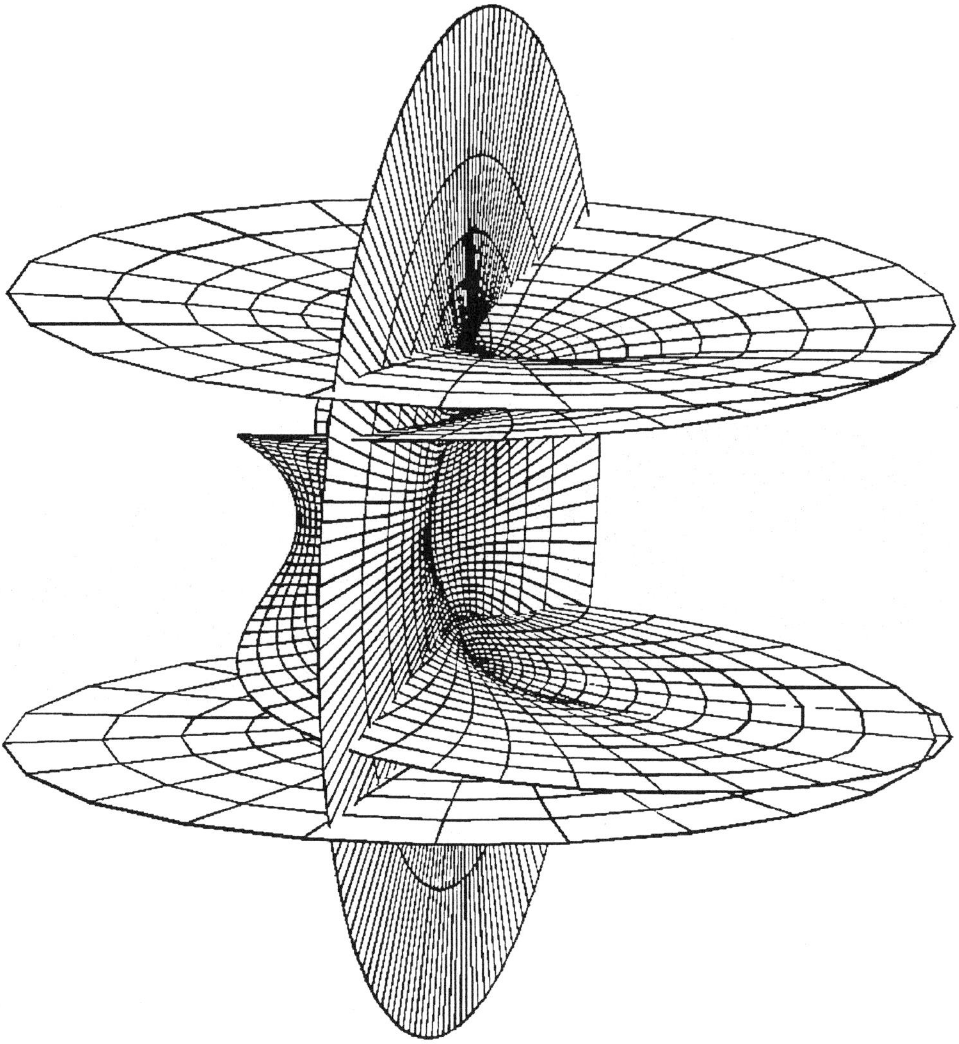

3. An immersed minimal surface of Enneper type with two asymptotic catenoidal ends and a vertical flat end (see page 43).

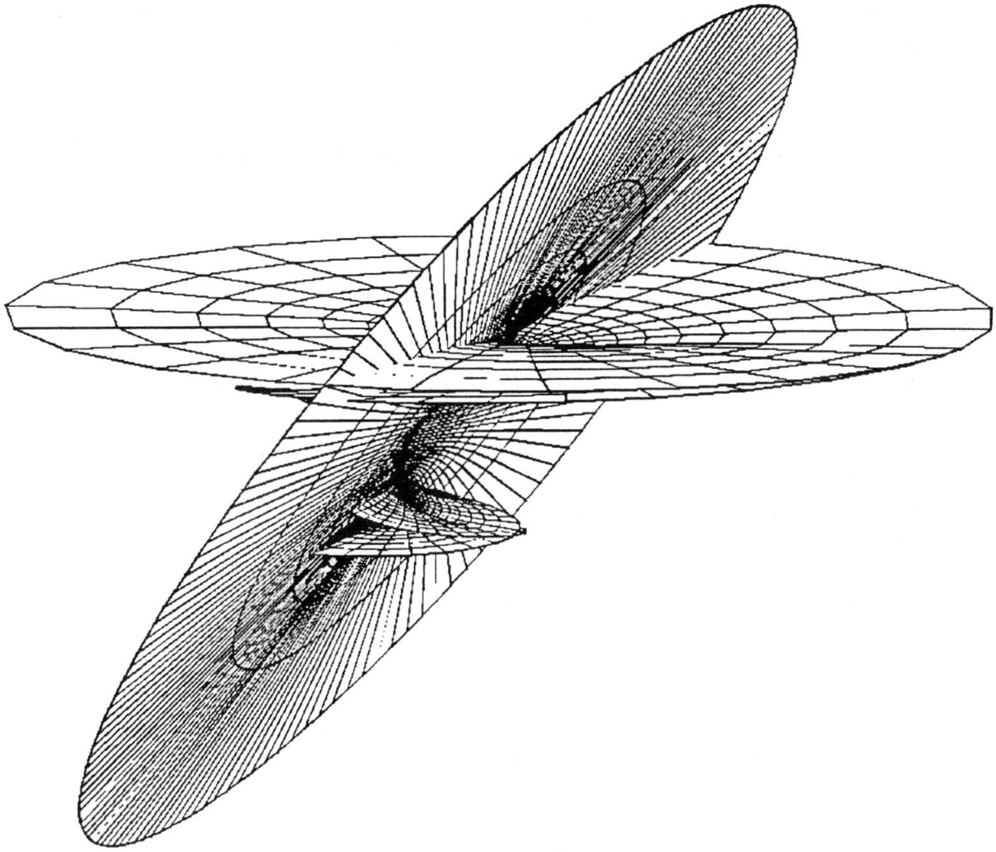

[889 by 796 pixels, 1.000 aspect ratio, 7.00 by 6.27 inches]

4. An immersed minimal surface of Enneper type where the flat end has been tilted (see page 44).

VII. BIBLIOGRAPHY

1) U. Abresch, "Constant Mean Curvature Tori in Terms of Elliptic Functions," Jour. Reine Angewan. Math. 374 (1987) 169-172.

2) A. I. Bobenko, "All Constant Mean Curvature Tori in R^3, S^3, H^3 in Terms of Theta-Functions," Preprint, Berlin 1990.

3) P. F. Byrd and M. D. Friedman, "Handbook of Elliptic Integrals for Engineers and Physicists," Springer 1953.

4) G. Darboux, "Lecons Sur la Theorie Generale des Surfaces, Vol. III, Gautier-Villars, Paris 1915 (Chelsea reprint).

5) H. Dobriner, "Die Flächen Constanter Krummung mit einem System Sphärischer Krummungslinien dargestellt mit Hilfe von Theta Functionen Zweier Variabeln, Acta Math. 9 (1886) 73-104.

6) H. Dobriner, "Die Minimal flachen mit einem Systen Sphärischer Krummungslinien," Acta. Math. 10 (1887) 145-152.

7) L. P. Eisenhart, "A Treatise on the Differential Geometry of Curves and Surfaces," Dover Reprint (1960).

8) A. Enneper, "Untersuchungen über die Flächen mit planen und sphärishcen Krummungslinien," Abh. Königl. Ges. Wissensch. Göttingen 23 (1878) and 24 (1880).

9) G. Fischer, "Mathematical Models" Friedr. Vieweg & Sohn, (1986). Commentary on Differential Geometry by M. P. doCarmo, G. Fischer, U. Pinkall, H. Reckziegel.

10) H. Hopf, "Differential Geometry in the Large," Lecture Notes in Math. 1000 (1983) Springer.

11) U. Pinkall and I. Sterling, "On the Classification of Constant Mean Curvature Tori," Annals of Math 130(1989) 407-451.

12) J. J. Stoker, "Differential Geometry," Wiley Interscience (1969).

13) R. Walter, "Explicit Examples to the H-Problem of Heinz Hopf," Geom. Dedicata 23, 187-213 (1987).

14) H. C. Wente, "Counterexample to a Conjecture of H. Hopf, Pacific Jour. of Math. 121,No. 1, 193-243 (1986).

Department of Mathematics
University of Toledo
Toledo, OH 43606

Editorial Information

To be published in the *Memoirs*, a paper must be correct, new, nontrivial, and significant. Further, it must be well written and of interest to a substantial number of mathematicians. Piecemeal results, such as an inconclusive step toward an unproved major theorem or a minor variation on a known result, are in general not acceptable for publication. *Transactions* Editors shall solicit and encourage publication of worthy papers. Papers appearing in *Memoirs* are generally longer than those appearing in *Transactions* with which it shares an editorial committee.

As of September 1, 1992, the backlog for this journal was approximately 9 volumes. This estimate is the result of dividing the number of manuscripts for this journal in the Providence office that have not yet gone to the printer on the above date by the average number of monographs per volume over the previous twelve months. (There are 6 volumes per year, each containing about 3 or 4 numbers.)

A Copyright Transfer Agreement is required before a paper will be published in this journal. By submitting a paper to this journal, authors certify that the manuscript has not been submitted to nor is it under consideration for publication by another journal, conference proceedings, or similar publication.

Information for Authors

Memoirs are printed by photo-offset from camera copy fully prepared by the author. This means that the finished book will look exactly like the copy submitted.

The paper must contain a *descriptive title* and an *abstract* that summarizes the article in language suitable for workers in the general field (algebra, analysis, etc.). The *descriptive title* should be short, but informative; useless or vague phrases such as "some remarks about" or "concerning" should be avoided. The *abstract* should be at least one complete sentence, and at most 300 words. Included with the footnotes to the paper, there should be the 1991 *Mathematics Subject Classification* representing the primary and secondary subjects of the article. This may be followed by a list of *key words and phrases* describing the subject matter of the article and taken from it. A list of the numbers may be found in the annual index of *Mathematical Reviews*, published with the December issue starting in 1990, as well as from the electronic service e-MATH [**telnet e-MATH.ams.org** (or **telnet 130.44.1.100**). Login and password are **e-math**]. For journal abbreviations used in bibliographies, see the list of serials in the latest *Mathematical Reviews* annual index. When the manuscript is submitted, authors should supply the editor with electronic addresses if available. These will be printed after the postal address at the end of each article.

Electronically-prepared manuscripts. The AMS encourages submission of electronically-prepared manuscripts in $\mathcal{A}\!\mathcal{M}\!\mathcal{S}$-TeX or $\mathcal{A}\!\mathcal{M}\!\mathcal{S}$-LaTeX. To this end, the Society has prepared "preprint" style files, specifically the amsppt style of $\mathcal{A}\!\mathcal{M}\!\mathcal{S}$-TeX and the amsart style of $\mathcal{A}\!\mathcal{M}\!\mathcal{S}$-LaTeX, which will simplify the work of authors and of the production staff. Those authors who make use of these style files from the beginning of the writing process will further reduce their own effort.

Guidelines for Preparing Electronic Manuscripts provide additional assistance and are available for use with either $\mathcal{A}_{\mathcal{M}}\mathcal{S}$-TEX or $\mathcal{A}_{\mathcal{M}}\mathcal{S}$-LATEX. Authors with FTP access may obtain these *Guidelines* from the Society's Internet node e-MATH.ams.org (130.44.1.100). For those without FTP access they can be obtained free of charge from the e-mail address guide-elec@math.ams.org (Internet) or from the Publications Department, P. O. Box 6248, Providence, RI 02940-6248. When requesting *Guidelines* please specify which version you want.

Electronic manuscripts should be sent to the Providence office only after the paper has been accepted for publication. Please send electronically prepared manuscript files via e-mail to pub-submit@math.ams.org (Internet) or on diskettes to the Publications Department address listed above. When submitting electronic manuscripts please be sure to include a message indicating in which publication the paper has been accepted.

For papers not prepared electronically, model paper may be obtained free of charge from the Editorial Department at the address below.

Two copies of the paper should be sent directly to the appropriate Editor and the author should keep one copy. At that time authors should indicate if the paper has been prepared using $\mathcal{A}_{\mathcal{M}}\mathcal{S}$-TEX or $\mathcal{A}_{\mathcal{M}}\mathcal{S}$-LATEX. The *Guide for Authors of Memoirs* gives detailed information on preparing papers for *Memoirs* and may be obtained free of charge from AMS, Editorial Department, P.O. Box 6248, Providence, RI 02940-6248. The *Manual for Authors of Mathematical Papers* should be consulted for symbols and style conventions. The *Manual* may be obtained free of charge from the e-mail address cust-serv@math.ams.org or from the Customer Services Department, at the address above.

Any inquiries concerning a paper that has been accepted for publication should be sent directly to the Editorial Department, American Mathematical Society, P. O. Box 6248, Providence, RI 02940-6248.

Recent Titles in This Series

(Continued from the front of this publication)

445 **S. Argyros, M. Lambrou, and W. E. Longstaff,** Atomic Boolean subspace lattices and applications to the theory of bases, 1991

444 **Haruo Tsukada,** String path integral realization of vertex operator algebras, 1991

443 **D. J. Benson and F. R. Cohen,** Mapping class groups of low genus and their cohomology, 1991

442 **Rodolfo H. Torres,** Boundedness results for operators with singular kernels on distribution spaces, 1991

441 **Gary M. Seitz,** Maximal subgroups of exceptional algebraic groups, 1991

440 **Bjorn Jawerth and Mario Milman,** Extrapolation theory with applications, 1991

439 **Brian Parshall and Jian-pan Wang,** Quantum linear groups, 1991

438 **Angelo Felice Lopez,** Noether-Lefschetz theory and the Picard group of projective surfaces, 1991

437 **Dennis A. Hejhal,** Regular b-groups, degenerating Riemann surfaces, and spectral theory, 1990

436 **J. E. Marsden, R. Montgomery, and T. Ratiu,** Reduction, symmetry, and phase mechanics, 1990

435 **Aloys Krieg,** Hecke algebras, 1990

434 **François Treves,** Homotopy formulas in the tangential Cauchy-Riemann complex, 1990

433 **Boris Youssin,** Newton polyhedra without coordinates Newton polyhedra of ideals, 1990

432 **M. W. Liebeck, C. E. Praeger, and J. Saxl,** The maximal factorizations of the finite simple groups and their automorphism groups, 1990

431 **Thomas G. Goodwillie,** A multiple disjunction lemma for smooth concordance embeddings, 1990

430 **G. M. Benkart, D. J. Britten, and F. W. Lemire,** Stability in modules for classical Lie algebras: A constructive approach, 1990

429 **Etsuko Bannai,** Positive definite unimodular lattices with trivial automorphism groups, 1990

428 **Loren N. Argabright and Jesús Gil de Lamadrid,** Almost periodic measures, 1990

427 **Tim D. Cochran,** Derivatives of links: Milnor's concordance invariants and Massey's products, 1990

426 **Jan Mycielski, Pavel Pudlák, and Alan S. Stern,** A lattice of chapters of mathematics: Interpretations between theorems, 1990

425 **Wiesław Pawłucki,** Points de Nash des ensembles sous-analytiques, 1990

424 **Yasuo Watatani,** Index for C^*-subalgebras, 1990

423 **Shek-Tung Wong,** The meromorphic continuation and functional equations of cuspidal Eisenstein series for maximal cuspidal subgroups, 1990

422 **A. Hinkkanen,** The structure of certain quasisymmetric groups, 1990

421 **H. W. Broer, G. B. Huitema, F. Takens, and B. L. J. Braaksma,** Unfoldings and bifurcations of quasi-periodic tori, 1990

420 **Tom Lindstrøm,** Brownian motion on nested fractals, 1990

419 **Ronald S. Irving,** A filtered category $\mathcal{O}_S$ and applications, 1989

418 **Hiroaki Hijikata, Arnold K. Pizer, and Thomas R. Shemanske,** The basis problem for modular forms on $\Gamma_0(N)$, 1989

417 **John B. Conway, Domingo A. Herrero, and Bernard B. Morrel,** Completing the Riesz-Dunford functional calculus, 1989

(See the AMS catalog for earlier titles)